T0137478

Innovation and Discovery in Russian Science and Engineering

Series Editors

Stavros Syngellakis
Wessex Inst of Technology, Southampton, Hampshire, UK

Jerome J. Connor
Massachusetts Institute of Technology, Cambridge, MA, USA

More information about this series at http://www.springer.com/series/15790

Nikolay N. Simakov

Liquid Spray from Nozzles

Experimental and Computer Simulation
of Hydrodynamics and Interphase Heat
and Mass Transfer

 Springer

Nikolay N. Simakov
Yaroslavl State Technical University
Yaroslavl, Russia

ISSN 2520-8047   ISSN 2520-8055   (electronic)
Innovation and Discovery in Russian Science and Engineering
ISBN 978-3-030-12448-9   ISBN 978-3-030-12446-5   (eBook)
https://doi.org/10.1007/978-3-030-12446-5

This Springer imprint is published by the registered company Springer Nature Switzerland AG.
The registered company address is: Gewerbestrasse 11, 6330 Cham, Switzerland

Preface

In this book, the new experimental facts about the features of two-phase flows formed by spraying of liquid in a gas by a nozzle are presented. Methods for calculating hydrodynamics and interphase heat and mass transfer in such flows are also presented.

The purpose of the book is to form a new view of the phenomenon that is widely applied in practice in technologies using spraying of liquids in a gas. New results of experimental studies and numerical simulation of hydrodynamics of an emerging two-phase flow and interphase heat and mass transfer in it are presented. The book could be interesting and useful for specialists who develop and use technologies with spraying of liquids in a gas, such as burning and pyrolysis of liquid hydrocarbons, granulation and drying of polymers, and dust and gas scrubbing.

As an author, I have been dealing with problems of modeling and calculating of spraying processes and devices for more than 30 years. During this time, two new physical phenomena were discovered. The results of my research were published in several journals (Russian and American ones) and in one book in Russian. The new book would be made available to a wider audience.

A set of unique features of this book include the following:

1. New experimental data of the phenomenon
2. Analysis of these data and new conclusions about the features of a strongly turbulent two-phase flow
3. Taking these features into account in the mathematical model of a two-phase flow
4. Numerical simulation and calculation of hydrodynamics and interphase mass transfer in a free two-phase flow
5. The same in a cylindrical apparatus

The book would benefit the reader since it presents new data on the phenomenon and the ways to take them into account in numerical calculations.

The book consists of 7 chapters, which include 104 images and 1 table.

Yaroslavl, Russia   Nikolay N. Simakov

Acknowledgments

The author is grateful to YSTU Associate Professor Yuri Grigorievich Zvezdin and Professor Boris Nikolayevich Basargin, who were initiators and initially leaders of experimental and computer research described in this book.

The author also expresses his deep gratitude to engineer Alexander Petrovich Plastinin, for his important contribution to the creation of new methods, the design of devices and their manufacture for pneumometric measurements, and participation in conducting experimental studies and obtaining the results which are described in Chap. 2 of this book.

Contents

Chapter 1
Introduction: Analysis of the Problems of Modeling of Hydrodynamics and Interphase Heat and Mass Exchange in the Processes with Spraying of Liquid in a Gas

Spraying of liquid in a gas, for example, using nozzles, is one of the three ways to increase the surface of phase contact and the intensity of interfacial heat and mass transfer processes. Spray processes are widely used in chemical technologies, energy, and transport. At the same time, theoretical methods for calculating such processes are not developed enough, which in turn does not allow reliably designing high-efficiency spray devices and is therefore a serious scientific problem.

1.1 Processes of Chemical Technology with Liquid Spraying

A number of heat and mass transfer processes of chemical technology such as rectification, absorption, drying, liquid fuel combustion and pyrolysis of liquid hydrocarbons, vapor condensation and evaporative cooling of liquids and high-temperature gases, wet removal of dust from industrial gas waste, etc. is carried out at the boundary of the liquid and gas phases. The efficiency of these processes is determined by the dynamics of the multiphase medium [1] and depends in a known manner on the specific surface area of the phase contact, which is the reason for the tendency to increase the latter in the carrying out of such processes [2].

To create a developed contact surface of gas and liquid phases, three main methods are currently used: (1) the formation of a liquid film flowing by contact with a gas, (2) the dispergation of the gas inside the liquid (bubbling), and (3) spraying the liquid into the gaseous medium with the help of various spraying devices. Each of these ways of creating a phase contact surface is answered by a special type of apparatus – these are, respectively, film columns, bubbling apparatuses, and hollow spray apparatuses [3]. The film columns, depending on the type of filler and the mode of their operation, make it possible to carry out both the first and second of these methods to increase the interfacial surface. In practice, combinations

N. N. Simakov, *Liquid Spray from Nozzles*, Innovation and Discovery in Russian Science and Engineering, https://doi.org/10.1007/978-3-030-12446-5_1

of various methods are also used, for example, spraying the reflux liquid in the top of the packed column to ensure an even distribution of liquid across the apparatus section.

In devices with an invariable phase contact surface [3] including hollow sprayers, for a given value of the specific surface area of the liquid phase, which in the case of liquid spraying is uniquely determined by the dispersion of the spray, the intensity of the heat and mass transfer process depends to a large extent on the relative velocity of the phases, which determines the magnitude of the coefficients of interphase heat and mass transfer. In other words, the efficiency of the chemical–technological process with the spraying of the liquid is due to the dispersion of the spray and the distribution of the liquid concentration and the phase velocities by the volume of the apparatus, i.e., hydrodynamic structure of the resulting two-phase disperse flow, called the spray.

Spraying of liquid, being one of the main ways of forming a developed interphase surface during various types of heat and mass transfer processes, is also used in other fields of technology, for example, in energy, in transport (internal combustion engines), and in a wide range of industries: chemical, metallurgical, machine-building, pulp and paper, food, etc.

Of the known methods for spraying liquids [4–11], mechanic and pneumatic spraying, carried out with the help of various types of injectors, were the most widespread in chemical technology [4, 7]. Mechanical injectors sometimes called hydraulic nozzles, to distinguish them from rotary mechanical sprayers, for example, rotating disk, and include jet, and centrifugal and jet-centrifugal nozzles. The latter provide a more even distribution of the liquid in the spray space, and by that they are sometimes called "nozzles of full spray" [12] and are most widely used in practice. In the spraying apparatus, one or more differently oriented nozzles are installed.

Apparatuses with liquid spraying, in particular "hollow spraying absorbers, are distinguished by their simple construction and low cost; they have low hydraulic drag and can be used with heavily polluted gases." However, up to the present time, "methods for calculating and designing hollow spraying absorbers have not developed, the influence of individual factors on their work is not sufficiently elucidated," which explains in part their "rather limited application" [12].

From the foregoing, it follows that the study of a two-phase disperse system, which is formed during a liquid atomization by a nozzle, including the development of a hydrodynamic model of a two-phase flow in a spray jet, is of undoubted interest for chemical technology and is an actual problem.

1.2 Hierarchy of Phenomena and Models of Complex Processes: Hydrodynamic Aspects of Modeling

The essence of the physical mechanism of any process of chemical technology is the transfer of mass, momentum, and energy from one area of space to another or from one part of the system to another part of it [13]. For this reason, the development of a theoretical model of a complex chemical–technological process is carried out on the basis of the theory of phenomena of transfer.

The application of the physical theory of transfer to the description of the processes of chemical technology is given a great deal of attention, as evidenced by the fact that along with the translated monograph [13], several monographs [14–18] were also published on this issue in Russian.

The main content of the books [13, 14] is the theoretical modeling of the transfer processes in a single-phase continuous medium. The first four chapters of [14] are devoted to the construction of a general closed system of transport equations in a single-phase multicomponent continuous medium on the basis of the laws of conservation of mass, momentum, and energy using the principles of linear nonequilibrium thermodynamics and applying this system of equations to solving specific problems of the transfer, mass impulse, and energy. Only in the last, fifth chapter are three specific examples of modeling the process of mass transfer in a two-phase medium. However, most mass exchange processes such as absorption, rectification, extraction, and drying are carried out in two-phase media; at that the heat and mass transfer takes place from one phase to the other through the interface [3].

Attempts to apply the methodology of the theoretical analysis of transfer phenomena in a single-phase medium to the case of two-phase media encountered a number of problems [14], the main one of which was the difficulty of "constructing a sufficiently general closed system of equations" describing transport phenomena in a multiphase medium. Two other problems are related to the difficulties of the physically correct "simplification of the mathematical formulation of problems" and the difficulties "in the formulation of physically justified boundary conditions." These problems caused the absence of "a unified theoretical basis for constructing physically rigorous models of transfer processes in multiphase media."

Nevertheless, the needs of practice stimulated the development of more or less common approaches to solving specific technical problems, which is the reason for the appearance of such works as [1, 14–18]. In particular, to date, we have accumulated a fairly rich experience in solving practical problems using the method of modeling transport processes in two-phase media, "based on the results of a theoretical analysis of the corresponding elementary transfer act" [14].

The essence of the method is as follows. First, the hydrodynamic pattern of the process is studied, and positions relating to the phases motion character are formulated. Then, on the basis of the hydrodynamic model, for a certain space-time region, balance equations are constructed, which are describing the intensity of the exchange by matter, by momentum, and by thermal energy between an individual particle of the dispersed phase and the surrounding continuous medium. At this one uses the results of solving the corresponding problem of an elementary transfer act in the framework of the transport theory for a single-phase medium.

An obvious and necessary feature of this approach is that at one of its stages the "hydrodynamic model of the process as the basis of a mathematical description" is studied [3]. The importance of this aspect cannot be overemphasized not only for theoretical but also for object modeling, since only "if there is a sufficiently correct description of hydrodynamic in qualitative and quantitative relation, it becomes possible to implement a large-scale transition" [3].

The hydrodynamic model of the process must describe the motion of phases in the apparatus at three levels of the "hierarchical structure of the physicochemical system"[3]: (1) the motion of an individual particle of the dispersed phase, (2) the constrained motion of the ensemble of particles of the disperse phase in the continuous phase (local hydrodynamics), and, finally, (3) the movement of phases throughout the entire volume of the apparatus, characterized by the structure of the flows in the apparatus. The description of hydrodynamics at the first mentioned level determines the mechanism of interphase transfer via the surface of an individual particle of the dispersed phase and, thus, the model of the kinetics of the elementary act of transfer. Summation (possibly with averaging [15, 16]) of the results of elementary acts of transfer for the ensemble of particles of the disperse phase in the elementary volume of the apparatus, taking into account local hydrodynamics, allows one to obtain a description of kinetic of interphase transfer in this elementary volume of the apparatus. Combining this description with the model of the structure of the phase flows in the apparatus provides the construction of a complete description of the transfer process. Such a complete description includes the characteristics of fields of speeds, concentrations, temperatures and distribution of the disperse phase, and hence of the interphase surface in the volume of the apparatus and their variation with time. It should be noted that macro local space-time changes of the above physical quantities, i.e., their changes in a volume element smaller than sizes of the apparatus but containing a sufficiently large number of particles of the dispersed phase, must be described by the model of the structure of the phase streams.

Thus, it is precisely the hydrodynamic model of the flows structure, reflecting the spatial distribution of velocities and concentrations of phases, that makes it possible to reduce the integration the equations of the kinetic of transfer with respect to time to integration over the spatial variable (over trajectory of particle) and obtain after additional integration over the flow cross section the macroscopic transfer effects for the whole apparatus as a whole.

At present, (1) models of complete (or ideal) displacement and (2) of complete (or ideal) mixing and (3) intermediate models [19] serve as typical cybernetic models of the structure of flows in volume apparatuses. Intermediate models in one way or another take into account the mixing of the flows due to diffusion, turbulence, or circulation of different scale and are represented by a cell model, one- and two-parameter diffusion models, and all sorts of combined models built on the basis of the simplest by adding to them stagnant zones, by introducing recirculation and bypassing [3, 19, 20].

The mathematical description of the process, with taking into account the actual mixing of the flows, becomes more complicated. The level of complexity of modeling increases to an even greater extent, if it is impossible to use for describing the flow structure for at least one of the two phases the simplest typical models, and one has to resort to models of partial mixing, including combined ones. The difficulties that arise in this case are related, first, to the uncertainty in the choice of these models themselves, because today there are no sufficiently valid recommendations for a single-valued choice of such models; second, with the known complexity of the procedure for experimental determination of the parameters of the model (number of cells, mixing coefficients, etc.) and verification of its adequacy [21]; and, third, with a significant complication of the mathematical description of the process by the necessity of introducing different kinds of efficiency coefficients and determining their values [3, 12, 21]. As a result, the calculation methods of processes and apparatus, taking into account the mixing and distribution of fluxes, are still insufficiently developed, and their application is limited in that in some cases there is no data on the necessary parameters, for example, longitudinal mixing coefficients [12].

Thus, the approach using typical models of the flow structure in describing the hydrodynamics of a combined process in a two-phase system has a number of significant drawbacks.

In view of the foregoing, in the construction of the hydrodynamic model of a two-phase system formed by spraying a liquid, it would be advisable to abandon the use of typical cybernetic models of the flow structure selected in some way a priori and use other more determinate concepts, for example, the fundamental laws of mechanics and the corresponding equations describing the motion of phases within the framework of the phenomenological approach considered below (Sect. 1.3.2).

1.3   Two-Phase Flow in a Jet Spray of a Nozzle

1.3.1   Modern Physical Concepts of the Spraying Liquid Process by Mechanical Injectors

A large number of works have devoted to spraying liquids with various devices, including injectors, in the chemical and other industries. Several monographs [4–11] reflected the results of numerous studies, the spray processes and devices, as well as the accumulated experience of their practical application. Perhaps the most significant subject in the content of these works is the spraying of liquids by the mechanical injectors of: jet, centrifugal and jet-centrifugal, which is due to the well-known advantages of sprayers of this type. Let's consider some specific aspects of this way of spraying liquids.

Upon the decay of the liquid flowing from the nozzle into the droplets, a so-called spray is formed, which is a stream of dispersed liquid in a gaseous medium. The most common cause of the liquid jet breakdown into droplets is the instability of the state of its continuous flowing. That in its turn is caused both by the prehistory of the emergence of this state, that is, the conditions of the flow of the liquid inside the injector, and by external factors affecting the liquid after its exit from the nozzle. The motion conditions of the liquid inside the injector include the value of excess pressure that determines the energy reserve of the fluid and its flow rate via the injector, as well as the shape and dimensions of the internal channels and diameter of the nozzle outlet hole. The nozzle design determines the configuration of the velocity field of the liquid flow and thus the shape and transverse dimensions of the jet or film flowing out of the nozzle outlet, its oscillation, and, in particular, the rotation of the liquid around the axis of jet.

External factors contributing to the crushing of the liquid flowing from the nozzle to the droplets include the hydrodynamic interaction of liquid with the gas and the action of gravity and other forces, for example, electrostatic ones. The physical properties of a liquid, such as surface tension, viscosity, and density, have a known effect on the spraying, in particular, on the dispersion of the spray.

When the liquid is sprayed with a centrifugal jet nozzle, the disintegration of the jet into droplets is basically completed at a distance of a few (up to ten) centimeters from the nozzle outlet hole [22]. The resulting droplet stream has a polydisperse composition, which in the future can change due to coalescence and secondary crushing of droplets during their collisions and interaction with the gas.

The cause of droplet collisions is the difference in the speeds of their movement. In addition, the probability of collisions depends on the concentration of droplets. After the disintegration of liquid flowing out the centrifugal jet nozzle and its bursting onto the primary droplets, the countable concentration n of droplets and, together with it, the volume fraction α of the liquid phase very rapidly and significantly decrease when removed from the nozzle. The changes in mean values over the spray flow cross section of these concentrations are approximately proportional to the square of the ratio of the nozzle outlet hole radius to the distance to it:

$$n \sim \alpha \cong (r/z \cdot \text{ctg}\,(\varphi))^2 \qquad (1.1)$$

Thus, at a distance $z = 100$ mm from a nozzle with a radius $r = 1$ mm and a root angle 2φ of sprayed jet, the volume fraction of the liquid α is of the order of 10^{-4} and further decreases even further. Along with it, as the distance from the nozzle increase, the probability of a collision of drops falls that can be estimated using the recommendations available in the literature [23–25]. However, even simpler estimates show that the probability of a droplet collision in the spray of a nozzle is relatively small of the order of several percent.

Determining the outcome of a collision of drops is much more difficult than to estimate its probability. Experimental studies, for example [26, 27], show that depending on the drop size, their relative velocity, and the physical properties of the liquid, an elastic collision, the merging of droplets, and crushing them into smaller droplets can occur.

A large number of papers, including [28–39], have been devoted to the investigation of the deformation and, as its consequence, of the decay of droplets during their interaction with a gas. That determines another mechanism of secondary crushing of the initial drops. However, the complexity of the phenomenon being considered, due to the presence of a large number of influencing factors and a wide range of their possible values, has not yet allowed to develop concrete and sufficiently general recommendations that allow one to unambiguously take into account the secondary crushing of droplets when calculating two-phase flows [40].

When processing experimental data on the decay of jets and droplets, the Weber criterion [6, 10] is considered as the main

$$\text{We} = \rho_{\text{g}} \cdot w^2 \cdot d/\sigma. \tag{1.2}$$

The relative velocity w of the mutual approach of droplets (or their movement in the gas flow), entering the criterion We, under the conditions prevailing in the spraying apparatus, usually does not exceed several tens of meters per second. In such a case, for most drops of the initial polydisperse stream, for which the diameters is $d < 10^{-4}$ m, the value of the criterion We is of the order of unity or less. In this case, the secondary crushing of the drops either does not take place at all or is insignificant [26–28, 35] and certainly should not lead to a significant change in the initial polydisperse composition of the drop stream formed due to the decay of the jet flowing from the nozzle. For this reason, in the modeling of processes and devices with atomization of a liquid by a nozzle, it is neglected for the secondary fragmentation of droplets, as, for example, in [10, 22]. It is necessary, however, to note that the data on a direct experimental verification of this assumption in the works mentioned are not given.

For the experimental substantiation of the possibility of neglecting the secondary fragmentation of droplets in the spray of the nozzle, it is necessary to determine the dispersed composition of droplets, characterized by the function of their distribution in size, in several cross sections of sprayed jet, if possible without interfering with the flow, and to analyze the droplet size spectrum dependence on the change in axial coordinate of the cross section. Typically, the dispersion of the sprayed liquid is measured in one cross section of the two-phase flow, most convenient for this.

The need for experimental measurements of the dispersion of the atomization is due to the fact that at present there are no unambiguous and satisfactory methods in the literature for calculating the disperse characteristics of spray of nozzles. The results of calculations based on formulas proposed by different authors do not agree well with each other, which is also noted in [22], and in some cases there is also a contradiction in the formulas proposed by the same authors for two bordering ranges of one of the defining parameters at the boundary junction these ranges, as, for example, in [10].

If in the modeling of the hydrodynamics of the spray flow there is information on the spectrum of droplet sizes that most fully reflects the dispersion properties of the liquid phase, then it can be used completely, by splitting the dispersed droplet composition into several monodisperse fractions, or partially – by replacing in the real model polydisperse flow of droplets by their monodisperse flow. As the common diameter of these droplets, the average diameter simultaneously on volume and area surface of the droplets of the polydisperse flow is usually chosen [10, 22, 41–47].

One of the most important aspects in the modeling of spraying processes is the hydrodynamic interaction of phases in a spray stream. In connection with this, it should be noted that, firstly, in accordance with the above remark on the smallness of the Weber We criterion, for most drops in the spray flow, their deformation by interaction with a gas can be neglected with sufficient accuracy [21, 48]. Secondly, because of the significant difference (approximately two orders of magnitude) in the dynamic viscosities of a liquid and a gas, the liquid circulation inside the droplets can be neglecting, treating them, thus, as solid spherical particles [21]. Thirdly, taking into account the smallness and rapid decrease in the volume fraction of the dispersed phase, with increasing the distance from the injector, often one neglects the mutual influence of the flow fields of the gas around individual particles [10, 21, 22, 48]. In this case, the boundary layers of individual particles do not overlap, and the strength of the interfacial interaction can be considered additive for the set of drops [21]. This assumption is considered valid for a volume fraction $\alpha < 5\%$ [48], which, in accordance with (1.1), is already satisfied at distances from the nozzle of the order of several of its diameters and is certainly satisfied where the crushing can be considered as finished ($z = 10–100$ mm) [22].

The total force acting on the droplet from the side of the gas flow includes the force of drag, Archimedes force, the force of attached mass, and also the force due to the presence of pressure gradients in the gas [49, 50]. Because of the low gas density, the three last components of the interfacial interaction force can be neglected [49]. In this case, the interfacial interaction is entirely due to the forces of drag or to the total drag of droplets. The drag force acting on a single drop moving in a fluid with relative speed W_{rel} is determined by the expression [50]:

$$F = 1/8 \cdot \pi \cdot d^2 \cdot \rho_g \cdot C_d \cdot (W_{rel})^2. \tag{1.3}$$

In case of a Stokes laminar flow about a particle (Re $< 0.1 \ll 1$), drag coefficient C_d can be found from the well-known and experimentally proven Stokes formula:

$$C_d = 24/\text{Re}. \tag{1.4}$$

Besides that for the drag coefficient C_d of solid spherical particles, there are several empirical relationships in the literature that are valid over a wide range of values of the Reynolds number Re, for example, it is the Klyachko formula, recommended in [35, 50] for $0 < \text{Re} < 700$

$$C_d = 24/\text{Re} + 4/\text{Re}^{1/3}, \tag{1.5}$$

or the formula of Rivkind and Ryskin [10, 22, 51] for $0 < \text{Re} < 200$

$$C_d = 1/(\mu_l/\mu_g + 1) \times \left[\mu_l/\mu_g \cdot \left(24/\text{Re} + 4/\text{Re}^{1/3}\right) + 14.9/\text{Re}^{0.73}\right],$$

which is more complicated in structure but gives for the liquid drops in the gas, practically the same results as the previous formula since in this case the ratio of viscosities $\mu_l/\mu_g \sim 10^2 \gg 1$.

Empirical formulas are also known for the volume density of the total force of interfacial interaction between a set of droplets and a gas. This, for example, the Ergan formula [52] used in [41–43], or more suitable for low-concentrated disperse flows and valid in a wide range of numbers Re, the modification of this formula [44]

$$dF/dV = f = \alpha \cdot \left(18 \times \mu_g/(d_{32})^2 + 0.36 \cdot \rho_g/d_{32} \cdot W_{rel}\right) \cdot W_{rel}, \qquad (1.6)$$

which gives for the drag coefficient of an individual particle the relation

$$C_d = 24/\text{Re} + 0.48. \qquad (1.7)$$

It should be noted that the listed (1.5), (1.7), and a number of other approximations of the experimental data for the drag coefficient of a solid spherical particle give in an interest range of numbers $0 < \text{Re} < 100$, although the close values, but differing in certain limits (Fig. 1.1).

In the hydrodynamic interaction of a sprayed liquid flow with a gaseous medium, a complex two-phase current arises, accompanied by a so-called ejection effect (influx, entrainment) of gas from the outer region into the spray space. Despite the abundance of works devoted to spraying of liquids by nozzles, this effect has not been fully studied to the present time, as evidenced by the existence in the modern literature of two conception used for physical interpretation and mathematical description of this phenomenon. According to one of them, based on the theory of turbulent jets [53], the gas ejection mechanism into the spray space is explained by

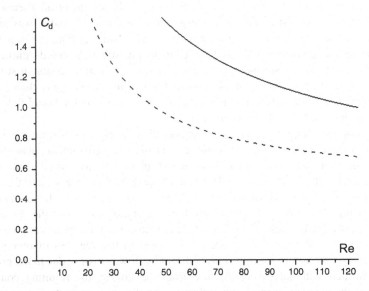

Fig. 1.1 Dependence of the drag coefficient of sphere on the Reynolds number; solid curve, calculation by formula (1.5); dashed curve, calculation by formula (1.7)

the interaction of turbulent moles of gas in the two-phase flow and its vicinity, which occurs in the absence of pressure gradients in the interaction region [54–58]. According to another, the inflow of gas from the external environment into the spray can be due to the presence of some pressure drop in the gas between the surrounding space and the spray region [49, 50].

The importance of this effect and the need to take it into account in the hydrodynamic structure of the spray stream is dictated by the fact that the inflow of gas into a spray is one of the main details of the general picture of a two-phase flow, in particular, the flowing of gas near the boundary of the spray. As the latter, it is possible to accept the surface, formed by the trajectories of droplets farthest from the nozzle axis [41], or the surface of the cone with an angle at the vertex equal to the root angle of the spray flow.

The gas inflowed into a free spray flow, when moving from its boundary to the axis, interacts with droplets, which should lead to curvature of their trajectories toward the axis of the stream and the redistribution of the fluid flow through its cross sections. In this case, the radial distributions of the specific density of liquid in the cross sections of the flow should vary on the height of the latter. These considerations are contradicted by the assumption told in [59] about the straightness of the trajectories of drops in the spray flow. From that the conclusion followed about the self-similarity of the radial profiles of the specific density of liquid in all cross sections of the spray flow. Moreover, on the basis of this assumption, some researchers have attempted to develop nozzles, "with the given form of the distribution of specific liquid density," including methods for calculating them [10, 60, 61]. It should be noted, however, that in the indicated studies there are no data on a direct experimental verification of the presence of the self-similarity of the normalized radial profiles of the specific liquid flows over the height of all stream. The measurement of the distribution of the dispersed phase in these studies was carried out only in one cross section of the spray flow and, in our opinion, not in the best way: the ring-shaped water collector [9] used in the investigations could significantly affect the spray flow, distorting its structure. That causes certain doubts first in the validity of attempts to ensure in all cross sections of the whole spray flow from a mechanical nozzle the same predetermined distribution of specific liquid flows and second in the reliability of the receiving results.

Based on the foregoing, it can be concluded that the process of spraying liquids with mechanical injectors is a very complex and still not well-studied phenomenon. The physical characteristics of the dynamic two-phase system, named as the spray flow, formed in this process, depend both on the design of the injector and the mode of operation (backpressure value) and on the hydrodynamic state of the gas medium in which the atomization is carried out: the gas, stationary before the injector is turned on, straight or back gas flow, a lateral gas flow around a spray stream, etc. In this case, due to the large difference ($\sim 10^3$ times) in the densities of phases, the nature of their motion in the root zone of the spray flow is mainly due to the nozzle. In the further, with increasing distance from the nozzle, the resulting complex motion of the phases is increasingly influenced by the gas outside the spray stream.

Based on these considerations, taking into account the possibility of neglecting the secondary fragmentation of drops, the entire problem of modeling the process of liquid spraying by a mechanical nozzle can be divided into three relatively independent tasks. This is, firstly, the internal hydrodynamic problem of the liquid flow inside the injector, the initial data for which are the design of the nozzle, the excess pressure of the liquid, and its physical properties. And as a result of the solution, the velocity field must be determined for the liquid flowing from the injector at the cross section of the outlet hole.

Secondly, it is a thermo- and hydrodynamic problem about the liquid jet and the formation of a drop stream in the root zone of a spray flow. The initial data for it are the solution results of the previous problem, that is, the distribution of velocities in the cross section of liquid jet at the nozzle outlet. In solving this, the second problem, it is necessary to determine the dispersion of the atomization, as well as the distribution of the concentrations and specific liquid flows in a cross section of the spray stream which is sufficiently close to a nozzle and where the crushing of liquid into droplets can be considered complete. Further we will call the said cross section the initial cross section of the zone of the formed spray flow or simply the initial cross section of the spray.

The third problem is the hydrodynamic problem about the two-phase flow of nozzle spray below its initial cross section and the flow of gas in the vicinity of the spray stream boundary. Initial data for it are the flow characteristics in the initial cross section of the spray, i.e., the results of solving the second problem, and the conditions for the flow of gas outside the cone of spray. Here, in essence, we are dealing with the problem of obtaining boundary conditions for a certain spatial region, which includes the basic zone of the spray flow.

To characterize the formulated problems on degree of their development in the scientific literature, it is necessary to note the following. The first of these is the problem of classical hydraulics, and the simplest formulation for the case of an "ideal centrifugal nozzle" [6] was first solved by G.N. Abramovich [62] with using the principle of maximum flow rate. Later similar results were obtained by the authors [63–65]. In a more general formulation for the case of a centrifugal jet nozzle, the problem becomes much more complicated, and at the present time there are no examples of its satisfactory solution in the scientific literature. The results given in [61] are valid only for a particular type of centrifugal jet nozzles – nozzles with axial and screw channels. In addition, it should be recognized that the use in this work for the determination the coefficients of the criteria equations of the aforementioned assumption about the similarity of the radial profiles of the specific liquid flows at the nozzle outlet hole and far from it, where they were actually measured, is not entirely correct, without sufficient justification for this.

The second problem in the formulation suggested is no less complex than, apparently, the absence in the scientific literature of examples of its theoretical solution is explained. An encouraging circumstance, however, is that in every separate case, it is possible to combine the first and second tasks into one and its solution, i.e., information on the structure of the flow of the dispersed phase in the initial cross section of the spray flow, one can obtain experimentally by means of known methods, which will be described below. At the same time, it is important to select the initial cross section of the spray flow correctly, satisfying simultaneously the requirements: (1) completion of crushing of the liquid by the time the jet reaches this cross section and (2) its relative proximity to the nozzle, so that the influence of the gas flow conditions in the outer region on the characteristics of the two-phase flow, in this cross section, was minimal. Examples of partial application of such an approach are known in the literature [22].

With regard to the third task, it is necessary to note first of all that in the suggested formulation, it is more general than the first two. Indeed, the statement of the problem, the methods, and, probably, some qualitative results of its solution are applicable not only to the case of spraying a liquid with mechanical but also pneumatic injectors and other sprayers. Examples of a more or less analogous approach to the modeling of the nozzle spray hydrodynamics are available in the literature; their consideration and detailed analysis are given in the following.

1.3.2   Phenomenological Models of Hydrodynamics of a Sprayed Liquid Flow

The theoretical model of hydrodynamics of a two-phase disperse system, formed when a liquid is sprayed into a gas medium, can be based on physical concepts that are different in content. A simulated two-phase medium can be considered (1) as a collection of two or more interacting and interpenetrating continua – a continuum approach [1, 49, 50, 66–68] – or (2) as a system consisting of a large number of interacting particles of two or more kinds, a discrete approach [69, 70]. A combination of these approaches is also used, when the dispersion phase is described as continuum and the dispersed phase as discrete [71]. At a discrete consideration, the macroscopic properties of the phases are determined as a result of averaging over the ensemble of particles.

In some cases, one can neglect the perturbation of the dispersion phase flow by the particles of a dispersed impurity, as is done, for example, in the study of atmospheric phenomena and in solving problems associated with pneumatic transport. In these cases, the task is simplified, because it reduces to determining the motion of particles of a dispersed phase in a known velocity field of the flow of dispersion phase [48]. However, in the modeling of two-phase flows formed during the dispersing of a liquid into a gas, such an allowance is not justified, since the masses of phases and their momentum flows are commensurable, i.e., values of the same order, which makes necessary to take into account the mutual hydrodynamic influence of the phases during their motion.

As for the characteristics of the models with respect to the solution methods, in most cases the equations for the mathematical description of two-phase systems from gas and droplets can be solved only by numerical methods with the use of computers. And if one or both phases are considered as discrete, then the calculation of the motion of a large number of particles causes a sharp increase of requirements to speed and memory volume of computer [50].

An important example of a phenomenological approach to the simulation of a two-phase system consisting of a gas and an atomized liquid which allows getting an analytic solution of the problem is the development of the theory of turbulent jets [53], taking into account the influence of the impurity of the dispersed phase [54–58].

The improvement of the semiempirical theory of turbulence of a gas jet is based on the self-similarity of a flow with an equilibrium turbulence structure [54]. The following assumptions are accepted:

1. The averaged local values of the velocities of the gas and droplets are assumed to be the same, and the pulsation velocities are different.
2. For commensurable masses of phases, the specific volume of the disperse phase is much less than the specific volume of the dispersion phase.
3. The linear size of heavy particles is many times smaller than the average linear size of turbulent moles, each of which is considered discrete from the moment of its formation to the moment of merging with the adjacent layer of the gas stream, which allows one to neglect the mutual influence of the particles of the dispersed impurity.

The perturbation of the turbulent gas jet by the admixture of the dispersed phase is taken into account in estimating the change in the pulsating velocity of the turbulent gas mole under the influence of the aerodynamic force of drag from the side of heavy particles using the law of conservation of the momentum of a turbulent mole containing impurity particles [55]. The theoretical estimate obtained in [54] is compared with the experimental data of [72], and their satisfactory agreement is established for shallow ($d < 30$ μm) particles.

Then, taking into account the first assumption, in the equations of averaged motion, the two-phase system is considered as a single continuous medium for which it is possible to obtain profiles of the averaged velocity and heavy impurity concentration by a manner which is analogous for the case of a single-phase jet. In this case, the Schmidt number for a two-phase medium is determined as the ratio of the turbulent transfer coefficients of momentum and the mass of impurity substance in the turbulent jet of the two-phase mixture [58]. Then the relation was derived relating this number to the impurity concentration and the Schmidt number for a single-phase gas mixture.

Further, in accordance with the adopted scheme of a two-phase turbulent flow, the differential equations of continuity and motion of the mixture are written as for a single medium in the boundary layer approximation. The same is done for the differential diffusion equation in which the impurity concentration is expressed by its mass fraction in the mixture and for the differential heat transfer equation for the case of low velocities in comparison with the speed of sound. After the integrating of these differential equations over jet cross section and some transformations [57], a system is obtained from six integral–differential equations describing the motion of a two-phase jet. The dynamic, diffusion, and thermal thicknesses of this jet, in general, are different.

In work [54] one compares the ratio of the thickness of the dynamic and diffusion boundary layers of a two-phase jet calculated within the framework of the given model and measured in the experiment [73]. This comparison indicates a satisfactory agreement between the calculated and experimental data.

In the case of a two-phase isothermal axisymmetric flooded jet without mass transfer of phases propagating in a gas without impurities, the mathematical description is simplified. In this case, a system of four equations is sufficient

$$\frac{d}{dx}\int_0^{\delta_u} y \cdot \rho \cdot U^2 dy = 0$$

$$\frac{d}{dx}\int_0^{\delta_u} y \cdot \rho \cdot U^3 \cdot dy = -2\int_0^{\delta_u} \rho \cdot \varepsilon \cdot \left(\frac{\partial U}{\partial y}\right)^2 \cdot y \cdot dy$$

$$\frac{d}{dx}\int_0^{\delta_\rho} y \cdot \rho \cdot U \cdot c \cdot dy = 0$$

$$\frac{d}{dx}\int_0^{\delta_\rho} y \cdot \rho \cdot U \cdot c^2 \cdot dy = -2\int_0^{\delta_\rho} \rho \cdot \frac{\varepsilon}{S_c} \cdot \left(\frac{\partial c}{\partial y}\right)^2 \cdot y \cdot dy, \qquad (1.8)$$

where x and y are longitudinal and transverse coordinates of the jet; δ_u and δ_ρ are dynamic and diffusion thickness of the jet; U, ρ, and $c(x, y)$ are speed, density of the mixture, and mass fraction of the impurity; and ε and S_c are the coefficient of kinematic turbulent viscosity and the Schmidt number of the mixture, respectively.

The system of equations (1.8) allows us to find four unknown functions in the initial section of the jet – the two boundaries $y_{1U}(x)$ and $y_{2U}(x)$ of the dynamic mixing zone and the two boundaries $y_{1\rho}(x)$ and $y_{2\rho}(x)$ of the diffusion mixing zone, and in the main section – the diffusion δ_ρ and dynamic δ_u boundaries of jet, as well as the laws of variation of the velocity $U_m(x)$ and the concentration $c_m(x)$ of impurities on the axis of the jet.

In practical calculations it is necessary to use dimensionless coordinates and expressions for the self-similar (along the axial coordinate of the jet) dimensionless velocity profiles and impurity concentrations in the main and initial parts of the jet, analogous to the case of a single-phase jet [74]. The required transformations of the system (1.8) and the calculation procedure are given in [56, 57]. In work [54] the results of calculating the isothermal two-phase jet from Eq. (1.8) are compared with the experimental results of the authors of [72]. That confirms a satisfactory agreement between the calculation and experiment.

Nevertheless, the approach developed by G.N. Abramovich with his students in works [53–58] has some limitations and for this reason cannot be recommended as a general basis for constructing a hydrodynamic model of a two-phase gas-droplet system formed when a liquid is dispersed in a gas medium. *These limitations include*, first, the assumption that the average value of the transverse velocity component is small compared with the averaged value of the axial component of the velocity for each phase in the jet [74]; second, the assumption of equality of the local averaged values of the phase velocity; and, third, the implicit assumption of the fact that a turbulent jet is created when the gas, which has a small specific volume of dispersed impurities, is flowing from some nozzle into the gas medium. The above assumptions and limitations are likely to be met with pneumatic dispersion of the liquid but are incorrect in the case of another dispersion method, for example, using a mechanical centrifugal jet nozzle.

Another disadvantage of this approach is related to the well-known condition that arises when modeling the flow in the apparatus, due to the need to take into account the interaction of the flow with the wall of the apparatus. And, apparently, the most important drawback of the method considered here is the impossibility, in setting up the problem of modeling a complex process, to divide it into two separate parts: modeling the hydrodynamics of the system and modeling the elementary acts of the transfer of heat and mass. Since the considered approach represents an inseparable combination of processes of turbulent exchange of mass, momentum, and thermal energy under the condition of self-similarity of these processes along the direction of the jet.

Another very common [1, 49, 66–68, 75] method of phenomenological modeling of two-phase flow consists in using the equations of classical hydrodynamics [76] for describing the motion of heterogeneous media in the approximation of interpenetrating continua.

The main problem of mathematical modeling of multiphase mixtures, as already noted, is in the construction of a closed system of equations of mixture motion for given physical and chemical properties of the phases and given mixture structure. In this case, it is usually necessary to introduce additional empirical or phenomenological relationships that take into account the non-one-phase nature, i.e., essentially different properties and behavior of phases (difference in densities, velocities, pressures, etc.) and their interaction with each other. These relations ensure the closure of the system of basic equations of motion, which express the laws of conservation of mass, momentum, and energy during the motion of a multiphase continuous medium.

The basic equations themselves can be derived in various ways [49]: by the phenomenological method, in the same way as in the case of a single-phase medium [76], or by a more detailed and explicit method of spatial averaging [49]. The difference between these methods is that the last makes it possible to obtain expressions for the macroscopic parameters of the phases, the stress tensors, the quantities that determine the interfacial interaction, and so on, and facilitates the choice of equations that close the system.

However, in deriving the basic equations of motion of a multiphase medium, a deeper method of spatial averaging is at the same time more complex. In addition, the "question of the choice of the averaging methods applied in this case is still controversial" [75]. For this reason, we confine ourselves here to the phenomenological approach in which a multiphase medium is regarded as a multi-velocity continuum consisting of several – in terms of the number of phases – interpenetrating continua.

This approach to modeling the motion of a two-phase medium is possible when the following assumptions are made:

1. The particle size of the dispersed phase in the mixture is much larger than the molecular-kinetic scales.
2. This dimension is much smaller than the distances at which the values of the macroscopic characteristics of the mixture or its individual phases vary substantially.

With these assumptions, it is possible to derive the equations of motion of the individual components (phases) and the mixture as a whole. The equations of the dynamics in the stresses and the energy equations for the individual phases are derived analogously to the case of a single-phase continuous medium with the only difference that the right-hand sides of these equations contain terms that take into account the interactions between phases including their mutual physical and chemical transformations.

We represent without deriving the basic equations describing the motion of the two phases of a disperse medium [49]. Taking into account the expression determining the operator of the substantial derivative for the i-th phase

$$\frac{d}{dt} = \frac{\partial}{\partial t} + \vec{V}_i \cdot \vec{\nabla} = \frac{\partial}{\partial t} + V_{ik} \cdot \frac{\partial}{\partial x_k}, \tag{1.9}$$

where the summation is over the coordinate index k, we have the equations for the mass

$$\frac{\partial \rho_i}{\partial t} + \vec{\nabla} \cdot (\rho_i \cdot \vec{V}_i) = J_{ji} \tag{1.10}$$

momentum

$$\rho_i \cdot \frac{d\vec{V}_i}{dt} = \nabla \sigma_i + \rho_i \cdot \vec{g}_i + \vec{P}_{ji} + J_{ji} \cdot \vec{V} \tag{1.11}$$

and energy of the i-th phase

$$\rho_i \cdot \frac{d}{dt}\left(U_i + \frac{V_i^2}{2}\right) = \vec{\nabla} \cdot (\vec{C}_i - \vec{q}_i) + \rho_i \cdot \vec{g}_i \cdot \vec{V}_i + E_{ji} - J_{ji} \cdot \left(U_i + \frac{V_i^2}{2}\right) \tag{1.12}$$

Here, in accordance with [49] denotes: ρ_i, V_i, U_i – density, velocity, internal energy, respectively, g_i – the stress of external mass forces, σ_i – stress tensor in the i-phase, C_i, q_i – quantities characterizing the action of external surface forces and heat influx, respectively, J_{ji}, P_{ji}, E_{ji} – the intensity of the transfer (from the j-th phase to the i-th) of mass, momentum and energy, respectively.

The average volume density of the i-th phase ρ_i, which appears in Eqs. (1.10–1.12), is related to the density ρ_{0i} of its substance by the relation

$$\rho_i = \rho_{0i} \cdot \alpha_i, \tag{1.13}$$

where α_i is the volume fraction of the i-th phase in the mixture.

For the closure of the system of basic Eqs. (1.10)–(1.12), it is necessary to obtain expressions for the stress tensors σ_i in the phases, for the quantities $J_{ji}, P_{ji}, E_{ji}, C_i$, and q_i characterizing interphase transfer, interaction, and transformation, through unknown variables α_i and V_i and the physical characteristics of the phases (density, viscosity, thermal conductivity, etc.). The form of these expressions can be determined on the basis of various considerations concerning the properties of a particular two-phase system and experimental data. In some cases, this is facilitated by an analysis of the statistical regularity characteristic of a given medium [15].

Specific examples of closed systems of equations of the form (1.10)–(1.12) can be the initial system of differential equations used to calculate the flowing near the end of a cylinder and a flat plate by a gas stream with hard particles [50] as well as the similar system of equations used for calculating the polydisperse flow of drops with a gas in the Laval nozzle [47]. In both models, the internal viscous friction in a gas is neglected, which makes it possible to simplify the problem.

To further simplify the hydrodynamic model of a two-phase flow, in addition to the above, the following assumptions are used. When calculating the hydrodynamics of a "cold" torch (with close and not too high temperatures of phases), it is sometimes possible to disregard the interfacial mass transfer by setting $J_{ji} = 0$. In this case, the continuity equations become homogeneous, and the corresponding terms of Eqs. (1.11) and (1.12) are equal to zero [50]. Taking into account the smallness of the relative volume of drops in the flare $\alpha_l \ll 1$ in Eqs. (1.10)–(1.12), it is possible to neglect the porosity difference from unity, setting $\alpha_g = 1 - \alpha_l \cong 1$. Usually the gas phase velocities are small compared with the speed of sound in the processes of chemical technology with liquid spraying. Then, it is permissible to neglect the gas compressibility [75, 76], considering its motion in a "cold" torch as incompressible fluid flow: $\rho_g = \rho_{0g} \cdot \alpha_g \cong \rho_{0g} = \text{const}$.

When describing the hydrodynamics of a free spray of a nozzle, it is often assumed that the gradients of the static pressure of the gas inside the flow and outside are zero [10, 22, 41].

Within the framework of the above simplifying assumptions, it is not necessary to involve the energy Eq. (1.12) in a system of equations for describing and calculating the hydrodynamics of the spray flow. For this it is sufficient to have the continuity Eq. (1.10), in this case homogeneous $J_{ji} = 0$, and the equations of momentum (1.11), in the right-hand parts of which apart from the mass forces $\rho_i \cdot g_i$ there are only the forces of interphase interaction P_{ji}. To approximate the latter, one can use expressions analogous to (1.6).

An example of practical application of the phenomenological approach under consideration to the investigation of the liquid spraying by nozzles is a (quasi) one-dimensional hydrodynamic model of the spraying liquid flow [41–43].

Assuming uniformity in the distribution of the hydrodynamic and thermophysical quantities over the cross section of spray flow, this model makes it possible to calculate their variation along its axial coordinate. The use of this model in the research of chemical and technological processes proved to be very fruitful: on its basis, methods for calculating heat and mass transfer processes such as the rectification and dust removal from gases in the spray "apparatuses of injector-type" [22, 46, 77, 78] and also high-temperature evaporation of hydrocarbon feed in the reactor for production of carbon black [45] and others.

One of the drawbacks of the model under consideration is its main property, (quasi) one dimensionality, which does not allow for taking into account the radial changes in the characteristics of the two-phase flow in the spray flow and beyond it, which in some cases (e.g., in the combustion and/or pyrolysis of liquid dispersed hydrocarbons, gas absorption by sprayed liquid, and the like) is highly desirable because of the significant impact of these changes on the average driving force of the process, the magnitude of heat and mass transfer coefficients, and, therefore, the efficiency of the process as a whole. When trying to eliminate the noted drawback using the quasi-two-dimensional "cell model of hydrodynamics of the free zone of the spray flow" [79, 80], difficulties arise. In particular, in this model, "the uncertainty of the initial and boundary conditions necessitated the introduction of two coefficients determining the initial velocity of the gas (*velocity in the initial cross-section, NS*) and the velocity of the gas at the boundary of the flow." These identification parameters were determined using experimental data, which to a certain extent, reduces the value of this model.

A comparison of the two above-described methods for modeling the hydrodynamics of the spray flow in the framework of the phenomenological approach, namely, the method based on the theory of turbulent jets [53–58] and the method based on the application of the classical hydrodynamics equations for the description of heterogeneous media in the continuum approximation [1, 47–50, 66–68, 75], allows us to distinguish the following common and distinguishing features of these methods.

The main general point is the fact that the mathematical descriptions of the two-phase flow are based on equations expressing the laws of conservation of mass, momentum, and energy of the disperse system. It is also general that in the equations of momentum, there are usually neglected terms describing internal physical friction in a gas phase [47, 50, 54].

The main difference is that in the first method the phase motions are considered as a joint one (their averaged velocities are equal to each other), and the change in the concentration of the dispersed phase and the temperature in the flow are described by differential equations of diffusion and heat transfer, the right-hand sides of which take into account the turbulent transfer of the dispersed phase and of heat in the jet

[54], while in the second method, the movement of phases occurs separately, generally speaking, with different velocities, which causes interphase exchange of impulses, heat, and mass transfer [47–50].

Another important difference is that according to the first method, turbulent momentum transfer takes place in the flow, which in the differential equation of momentum is taken into account by the term describing turbulent friction [54]. In the equations of motion of the gas corresponding to the second method of description, analogous terms are absent [47, 50]. Another difference is due to the neglect in the first method by the static pressure gradient and its account in the second method.

Some special differences between the methods are related to (1) the set of equations forming a closed system, e.g., in the first method, the equation of motion is used only for the axial velocity component; (2) with the form of the equations, only stationary in the first method, and, possibly, nonstationary in the second; (3) with the type of dependent variables, e.g., in the first method the equations are written relative to the averaged velocity components and in the second method with respect to the total velocities; and (4) with methods of solving systems of equations – analytical, numerical, etc.

The listed differences of the considered models cause a possible difference in the results of calculations obtained with their help. At the same time in the literature, there are no recommendations for unambiguous choice of this or that model in the application to the description and calculation of the spray hydrodynamics of a nozzle. A comprehensive experimental study of the hydrodynamic structure of the spray flow should help to solve the problem of model selection, designed to reveal the most characteristic aspects of the phenomenon, requiring their reflection in the mathematical description.

Thus, currently the methods for calculating the hydrodynamic structure of a two-phase flow in an axisymmetric spray of a mechanical nozzle based on a two-dimensional (and even more so three-dimensional) model constructed within the framework of the phenomenological approach are not well developed and presented in the Russian literature on the chemical technology, which hampers the development of effective methods for using these processes and their corresponding equipment.

1.4 Experimental Research Methods of Hydrodynamics of Two-Phase Flows

To substantiate the physical model and construct the mathematical description of the hydrodynamics of a two-phase flow formed during the spraying of a liquid, in particular, to set the initial and boundary conditions of a mathematical model, and also to verify its adequacy, it is necessary to have experimental data of the measurement of the main characteristics of the hydrodynamic structure of the spray stream of nozzle, which include the distributions of velocities $V_1(r, z)$, concentrations $\alpha_1(r, z)$

and specific liquid flows $j_l(r, z)$, the dispersion characteristics of liquid phase $f(d)$ and d_{32}, as well as the distribution of gas velocity $V_g(r, z)$ and its static pressure $P(r, z)$.

Let us consider some of the existing methods for measuring the hydrodynamic characteristics of a two-phase flow separately for dispersed and continuous phase.

Among the known methods [61–84] of measuring the velocity, concentration and dispersion of liquid particles and contactless methods are most preferable, enabling one to obtain information without mechanical interference in a flow that disturbs its structure. These include a variety of optical methods: photographic [81], in particular, colors two-pulse photography [22] and high-speed filming [85], holographic [86], and methods based on laser scattering. The latter include laser Doppler [87] and time-of-flight [88–91] anemometry as methods for determining the velocity of particles and a method for the small-angle light scattering indicatrix [92–94] to determine their dispersion and concentration.

Laser light scattering methods have greater operational capabilities in comparison with photographic and holographic methods, since they are not related to the need for processing photographic images; they make it possible to automate the processing of the light scattering signal on particles and increased speed of measurement.

Determination of the particle velocity by using laser anemometry involves the creation in the measuring zone of a nonuniform in the direction of particle flight of a light field of laser radiation, the optical extraction of a pulsed light scattering signal from a particle, and the processing of this signal with measurement of its times parameter uniquely related to the particle velocity. In Doppler anemometry [87] as such time parameter is used the Doppler shift of the radiation frequency, and in the time-of-flight anemometry [88–91] it is the time of flight by the particle of the base distance.

Currently, for measuring the particle velocity, two versions of the time-of-flight method are used, in which the basic distance is specified, respectively, by means of two laser beams [88, 89] or one laser beam of Gaussian type [90, 91]. The time-of-flight method with one beam, compared with two-beam method and Doppler anemometry requires a smaller and simpler set of optical instruments for forming the light field and separating the light scattering signal. This simplifies the alignment of the optical circuit and the operation of the device as a whole and thereby increases the reliability of measurements.

Of the existing methods [84] for measuring the dispersity of aerosols in the spray nozzle of nozzles, the abovementioned method of small-angle light scattering, which has been widely used in recent years, seems to be the most promising. The basic principles of the small angles method (SAM), formulated by K.S. Shifrin and co-authors [92, 93], will be discussed in detail in the next chapter. Here we note that a number of examples of practical applications of SAM and an extensive bibliography by the method are given in [94]. Note that the SAM makes it possible to determine not only the dispersion of [95] but also the average concentration of the dispersed phase [93, 96] in the scattering volume. The appropriate choice of configuration, dimensions and spatial arrangement of this scattering volume, makes it possible to measure the liquid distribution in the spray flow.

One of the alternatives to noncontact optical methods for determining the dispersion of a spray is the frequently used [9, 10, 22] method of trapping droplets by an immersion medium, in particular by an oil layer, followed by microphotography of a sample, measuring droplet sizes by their images on a photograph, and statistical processing of the results of these measurements. However, this method is less preferable because of its drawbacks, such as a violation of the flow structure when a drop catcher is introduced into it and a disturbance in the measured disperse structure of the aerosol due to crushing and merging of droplets during sampling, and, finally, because of the complexity and laboriousness of the measurement procedure.

The distribution of the specific liquid fluxes can be determined by calculation from experimental data on the distribution of velocities and concentrations of the dispersed phase, or simpler by direct measurements with the use of simple water collectors in the form of probe tubes. At this it is necessary to take care of providing of the minimal perturbing influence of the probe moisture collector on a two-phase stream.

For this it is necessary that the probe has transverse dimensions relatively small in comparison with the radius of the spray flow. In addition, it is highly desirable to ensure the selection of liquids by performing the so-called isokinetic flow conditions, i.e., the equality of the gas velocity at the entrance to the probe cavity and its velocity at the same "point" of the unperturbed flow in the absence of the probe. At small dimensions of the probe tube, gas braking, even in the absence of its artificial suction to the cavity of the probe, insignificantly affects the result of measuring the specific liquid flows, as evidenced by both the calculated estimations of droplet deceleration corresponding to these conditions as well as the results of direct experiments. This, of course, does not truly in the case of using large sectionalized circular horizontal, vertical [5, 7] and similar "universal samplers" [10].

Measurement of the hydrodynamic characteristics of the gas phase of a spray stream with the help of the aforementioned noncontact optical methods proves to be very problematic on the one hand due to the optical transparency of the gas and, on the other, due to the presence of a dispersed phase. Indeed, to study the gas motion outside the spray stream using optical methods, the visualization of the gas flow by introducing a tracer into it, in particular smoke particles [97], can be successfully used, with further photographing the flow pattern, but inside the spray stream, this is absolutely impossible due to a violation laminar flow as a result of turbulence and the creation of a strong optical background by the flow of a dispersed phase.

The latter reason does not allow using also the methods of laser Doppler or time-of-flight anemometry. The problem of using noncontact methods for studying the gas flow in this case is further complicated by the fact that in addition to the distribution of the gas velocities, it is necessary to measure the distribution of its static pressures in the spray stream and in its vicinity.

Thus, we inevitably arrive at the use of probe methods for these purposes [98]. However, in this case, for example, when measuring the gas velocity with a thermo-anemometer, care should be taken to protect the sensor against the dispersed phase of the flow [81] and prevent its influence on the results of measurements of the characteristics of the gas flow. From the point of view of the unity of the experimental methods for determining the velocity and pressure of the gas in the spray flow, there is very suitable well-known pneumometric method [9], which is widely used in experimental gas dynamics. It is based on the use of various types of receivers of total and static gas pressure in a stream, e.g., cylindrical, spherical probes, or Pitot-Prandtl tubes [98–100], in conjunction with a sufficiently sensitive micromanometer [101]. Of course, it is also necessary to take appropriate measures to eliminate the negative effect of the disperse phase on the measurement procedure.

1.5   Kinetics of Elementary Acts of Heat and Mass Transfer

Depending on phase (continuous or dispersed) which limits the heat and mass transfer, three problems are distinguished connected with heat and mass transfer in a two-phase disperse system [21]. The first is an external task with a boundary condition of the second kind determining the flow of heat or mass through the interface of phases. The second is internal task with a boundary condition of the first kind, setting the temperature or concentration on the boundary surface. And the third is combined (or mixed) task with a boundary condition of the third kind, equalizing the external and internal fluxes at the boundary of phases.

When considering the heat transfer (without taking into account evaporation) between a single drop with the temperature on the surface T_s, and the gas surrounding it with temperature T_g, the heat flux (per unit area of the droplet surface) is determined by using the equation of heat transfer [21]:

$$q = \frac{1}{S} \cdot \frac{dQ}{dt} = \alpha \cdot (T_s - T_g). \tag{1.14}$$

The coefficient α of convective heat transfer can be determined from the well-known [83] approximation for the Nusselt number

$$\alpha \cdot d/\lambda_g = \mathrm{Nu} = 2 + 0.03 \cdot \mathrm{Re}^{0.54} \cdot \mathrm{Pr}^{0.33} + 0.35 \cdot \mathrm{Re}^{0.58} \cdot \mathrm{Pr}^{0.36} \tag{1.15}$$

or by another way [102], more simple and convenient,

$$\mathrm{Nu} = 2 + 0.6 \cdot \mathrm{Re}^{1/2}\mathrm{Pr}^{1/3}. \tag{1.16}$$

In the sprayed flow from the nozzle for droplets with an average diameter $d = 2R \approx 0.15$ mm the convection inside them can be neglected, treating them as hard balls [21]. Then the temperature on the surface of the drop $T_s = T(R, t)$ and also at any internal point $T(r, t)$ can be found by solving the mixed problem of heat exchange, i.e., heat equation

$$\rho \cdot c \cdot \partial T/\partial t = \mathrm{div}(\lambda \cdot \mathrm{grad} T)$$

or more simple equation

$$\partial T / \partial t = a \cdot \Delta T \tag{1.17}$$

at constant thermal conductivity (λ = const) of the droplet substance and the temperature conductivity $a = \lambda/(\rho \cdot c)$ with the boundary condition of the third kind

$$-\lambda \cdot \partial T(R, t)/\partial r = \alpha \cdot (T_s - T_g). \tag{1.18}$$

under the initial condition

$$T(r, 0) = T_0. \tag{1.19}$$

For the case T_g = const, in points far from the drop surface, the solution of the problem (1.17–1.19) is known [83]. It depends on the dimensionless parameter Bi = $\alpha \cdot R/\lambda$ – the Bio number. And in the dimensionless variables $\xi = r/R$, Fo = $a \cdot t/R^2$ (Fourier number), and $\theta(\xi, \text{Fo}) = (T_0 - T_g)/(T_0 - T_g)$, it can be represented as a series of own (eigen) functions

$$\theta = \sum_{n=1}^{\infty} A(\mu_n) \cdot U(\mu_n \cdot \xi) \cdot e^{-\mu_n^2 \cdot \text{Fo}}, \tag{1.20}$$

where

$$A(\mu_n) = \frac{2 \cdot (\sin \mu_n - \mu_n \cdot \cos \mu_n)}{\mu_n - \sin \mu_n \cdot \cos \mu_n}, \quad U(x) = \frac{\sin x}{x}, \tag{1.21}$$

and the own (eigen) values μ_n are solutions of equation

$$\text{tg}\, \mu_n = \frac{\mu_n}{1 - \text{Bi}}. \tag{1.22}$$

For Fo > 0.3, the series (1.20) converge rapidly, and with accuracy 1% we can confine ourselves to only one of its first terms.

For drops of water in the air (in a spray flow) under usual conditions, we obtain the following estimates: Pr = 0.71; Re = 40–120; Nu = 7 ± 1.5; λ_g = 0.026 W/(m·K); α = 1300; Bi = 0.3; a = $1.4 \cdot 10^{-7}$ m²/s; $u_z \approx$ 20 m/s; Fo = 25 $t \approx 25z/u_z \approx z/0.8$; and Fo > 0.3 at z > 0.24 m. Thus, at the distances z from the nozzle larger than 0.24 m, in the relations (1.20–1.22) one can confine ourselves to only one first term of the series with μ_1 = 0.92. In a region closer to the nozzle, i.e., in the active zone, this approximation is not enough.

In the active zone (z < 0.3 m), the residence time of drops in the gas flow $t = z/uz <$ 0.015 s and the Fourier number Fo < 0.38 are small; in solution (1.20–1.22) it needs to take into account several first terms of series, which complicate the calculations.

On the other hand, while the droplets are in the region close to the nozzle, their temperature changes only in a thin near-surface layer and does not have time to change substantially in the nucleus of the droplet. In this case, the problem of the propagation of heat in this layer is analogous to the one-dimensional heat conduction problem in a semi-bounded body [83] (or on a semi-bounded straight line [103]). For a boundary condition of the first kind on the surface of the drop $T(0, t) = \text{const} = T_s$ (the coordinate x is measured from the surface of the drop inward) and for the initial condition $T(x, 0) = T_0$, the solution of this task is known and represented in terms of the error integral:

$$\frac{T(x,t) - T_0}{T_s - T_0} = 1 - erf\left(\frac{x}{2 \cdot \sqrt{a \cdot t}}\right). \tag{1.23}$$

The density of the heat flux on the drop surface in this case changes as

$$q = \frac{\lambda \cdot (T_s - T_0)}{\sqrt{\pi \cdot a \cdot t}} \tag{1.24}$$

Note that the solution of this problem is known even under boundary conditions of the second kind, and with a phase transition inside the layer of matter [83], it is not difficult to find its solutions also for boundary conditions of the third kind. Analytical solutions of a number of heat exchange problems for a plane layer and a sphere were obtained in the work [104].

However, in all the cases considered above, the drawback of the available solutions is that they use the conditions of the constancy (1) of the gas temperature T_g and the heat transfer coefficient α (and, together with the latter, the relative velocity of the droplet and gas) or (2) of the temperature at the surface of the drop T_s or (3) of the heat flux density $q(t)$ through the surface of the drop. In general, none of these conditions is satisfied in the spray flow. Moreover, the dependencies of said quantities on time are not known in advance, which makes it impossible to construct an analytical solution of the problem of heat exchange of a gas with a droplet moving in it.

In contrast to the analytic solutions considered above, numerical methods for solving the heat equation with various boundary conditions, including those of the third or second kind, which are of interest to us, are free of these limitations and shortcomings caused by them. Therefore, numerical methods can be recommended as a suitable alternative to analytical methods.

In modeling the mass exchange process between a single drop and a gas, in the case when only the physical absorption of a gas component occurs by the drop surface and subsequent diffusion of this component in the drop, according to the similarity theory, the mathematical description of the mass transfer will be exactly the same (up to the notation of the quantities), as for heat exchange, and everything that has been said about the modeling of heat exchange can be transferred to mass exchange. The above and a number of other problems of the mass exchange with their solutions are set forth in the remarkable monograph [105].

In the presence of a chemical reaction between the gaseous component which is absorbed and diffusing in a droplet and the substance of droplet, a phenomenon of chemisorptions occurs, for example, at the wet purification of industrial gaseous emissions. Models of chemisorptions kinetics are developed and presented in the scientific and technical literature insufficiently. Therefore, in this work it is desirable to give them attention.

1.6 Conclusions on Part 1: Formulation of the Research Tasks

Based on the above analysis of the current state of the issues related to the methods of modeling and calculating processes with liquid spraying, one can arrive at the following conclusions.

One of the reasons restraining the creation of new ones, as well as the development and increase of efficiency of existing spraying processes and apparatus, is the lack of sufficiently general, strictly justified, accurate, and reliable methods for calculating these processes.

The physical scheme underlying the mathematical description of the hydrodynamics of a two-phase flow created by an injector requires experimental refinement, in particular, regarding the existence and role of secondary crushing of liquid droplets, the difference or coincidence of phase velocities, and the presence and magnitude of static gas pressure gradients in the spray flow. The clarification of the latter question is necessary both for concretization, in particular, for simplifying the mathematical description of the hydrodynamic structure of the spray stream in the case when the gas pressure gradients can be neglected, and for elucidating the mechanism of ejection (suction) of gas from the outer region into the spray flow.

As a phenomenological model of the hydrodynamics of a liquid spray, it is advisable to consider each of the phases as a quasi-continuous medium, to use the classical equations of the mechanics of continuous media: the equation of continuity, momentum, and energy. In this case, the general equations of the mathematical model (1.10–1.12) must be concretized in accordance with the physical scheme chosen on the basis of experimental research and supplemented by relations closing the equations system.

For the constructed in this way mathematical model of the hydrodynamics of a two-phase flow, it is necessary to formulate the initial data, i.e., the initial and boundary conditions, to select and justify the calculation methods for more or less general cases, and to verify the adequacy of the model by comparing the calculation results with the experimental data.

The problems associated with modeling the kinetics of elementary transfer acts, in particular, the chemisorption kinetics, also require refinement and additional studies.

In accordance with the above analysis of the problem of modeling hydrodynamics and interphase heat and mass exchange in the processes with spraying of liquid in a gas, the objectives of this study are stated as this way:

1. Experimental study of the hydrodynamic structure of a two-phase flow in a free spray created by a mechanical centrifugal jet nozzle, including the measurement of spatial distributions of the main hydrodynamic characteristics of the phases: velocity and static pressure of gas and dispersion, velocity, concentration, and specific flows of liquid
2. Analysis of the results of the experiment and the selection on the basis of this analysis of the physical model of a two-phase system in a free spray flow
3. Development of a mathematical description and methods for calculating the hydrodynamics of the spray flow from a mechanical injector
4. Study of the kinetics of elementary acts of heat–mass transfer at the level of an individual particle of a dispersed phase
5. Development of mathematical models of complex spray processes that combine the hydrodynamics of a two-phase flow with the kinetic of an elementary act
6. Practical application of these models to the calculation of some processes with liquid spray

References

1. Nigmatulin, R. I. (1987). *Dynamics of multiphase media, part 1*. Moscow: Nauka.
2. Kasatkin, A. G. (1973). *Basic processes and apparatuses of chemical technology*. Moscow: Khimiya.
3. Kafarov, V. V. (1979). *Fundamentals of mass transfer*. Moscow: Vysshaya Shkola.
4. Wittmann, A. A., Katznelson, B. D., & Paleev, I. I. (1962). *Spraying liquid nozzles*. Moscow: Gosenergoizdat.
5. Borodin, V. A., et al. (1967). *Sprays of liquids*. Moscow: Mashinostroenie.
6. Dityakin, Y. F., et al. (1977). *Sprays of liquids*. Moscow: Mashinostroenie.
7. Pazhi, D. G., Prakhov, A. M., & Ravikovich, B. B. (1971). *Nozzles in the chemical industry*. Moscow: Khimiya.
8. Pazhi, D. G., Koryagin, A. A., & Lamm, E. L. (1975). *Spraying devices in the chemical industry*. Moscow: Khimiya.
9. Pazhi, D. G., & Galustov, V. S. (1979). *Sprays of liquid* (p. 216). Moscow: Khimiya.
10. Pazhi, D. G., & Galustov, V. S. (1984). *Fundamentals of spraying technology*. Moscow: Khimiya.
11. Golovachevsky, Y. A. (1974). *Sprinklers and nozzles of scrubbers of the chemical industry*. Moscow: Mashinostroenie.
12. Ramm, V. M. (1976). *Absorption of gases*. Moscow: Khimiya.
13. Bird, R., et al. (1960). *Transport phenomena*. New York: Wiley. Khimiya, Moscow, 1974.
14. Protodyakonov, I. O., Marculevich, M. A., & Markov, A. V. (1981). *Phenomena of transfer in the processes of chemical technology* (p. 263). Moscow: Khimiya.
15. Protodyakonov, I. O., & Chesnokov, Y. G. (1982). *Hydromechanics of the fluidized bed*. Leningrad: Khimiya.
16. Protodyakonov, I. O., & Bogdanov, S. R. (1983). *Statistical theory of transfer phenomena in the processes of chemical technology*. Leningrad: Chemia.

17. Protodyakonov, I. O., & Syshchikov, Y. V. (1983). *Turbulence in the processes of chemical technology*. Leningrad: Nauka.
18. Goldshtik, M. A. (1984). *Transfer processes in a granular layer*. Novosibirsk: ITF SB of the USSR Academy of Sciences.
19. Romankov, P. G., & Kurochkina, M. A. (1982). *Hydro-mechanical processes of chemical technology*. Leningrad: Khimiya.
20. Zhorov, Y. M. (1978). *Modeling of physical and chemical processes of oil refining and petrochemistry*. Moscow: Khimiya.
21. Brounshtein, B. I., & Fishbein, G. A. (1977). *Hydrodynamics, mass and heat transfer in disperse systems*. Leningrad: Khimiya.
22. Aniskin, S. V. Dissertation, LTI CBP, Leningrad.
23. Babukha, G. L., & Shrayber, A. A. (1972). *Interaction of particles of a polydisperse material in two-phase flows*. Kiev: Naukova Dumka.
24. Gatsev, V. A., et al. (1974). Collision of particles in mutually perpendicular flows of sprayers of chemical technology. In *Chemical technology* (pp. 67–70). Yaroslavl: Yaroslavl Polytechnic Institute.
25. Gatsev, V. A., et al. (1974). On the collision of particles in the sputtered streams of chemical engineering spraying machines. In *Chemical technology* (pp. 71–76). Yaroslavl: Yaroslavl Polytechnic Institute.
26. Arkhipov, B. A., et al. (1978). Experimental study of droplets interaction in collisions. *Zhurnal Prikladnoi Mekhaniki i Tekhnicheskoi Fiziki (Journal of Appleid Mechanics and Technical. Physics)*, (2), 21–24.
27. Babukha, G. L. (1972). Experimental study of the stability of droplets in collisions. In *Teplofizika i teplotekhnika (Thermo physics and heat engineering)* (Vol. 21, pp. 89–96). Kiev: Naukova Dumka.
28. Borodin, V. A., et al. (1982). On the fragmentation of a spherical droplet in a gas stream. *Zhurnal Prikladnoi Mekhaniki i Tekhnicheskoi Fiziki (Journal of Applied Mechanics Technical Physics)*, (1), 65–92.
29. Volynsky, M. S., & Lipatov, A. S. (1970). Deformation and fragmentation of drops in the gas flow. *Inzhenerno-Fizicheskiy Zhurnal (Engineering and Physics Journal)*, 18(5), 838–843.
30. Honor, A. L. (1978). Movement and spreading of a drop in the gas flow. In *Some questions of mechanics of continuous media* (pp. 173–187). Moscow.
31. Honor, A. L., & Zolotova, N. V. (1981). Braking and deformation of a liquid drop in a gas stream, Izvestiya AN SSSR. *Mekh. Zhidkosti i Gaza (Mechanics of Liquid and Gas)*, (2), 58–69.
32. Honor, A. L., & Zolotova, N. V. (1981). Decay of the drop in the gas flow. In *Gas dynamics of nonequilibrium processes* (pp. 42–45). Novosibirsk: Institute of Theoretical and Applied Mechanics of SB AS USSR.
33. Klyachko, L. A. (1983). To the theory of fragmentation of a drop by a gas flow. *Inzhenerno-Fizicheskiy Zhurnal (Engineering and Physics Journal)*, 3(3), 544–557.
34. Simons. (1976). Acceleration and deformation of a liquid drop. *Raketn. Tekhn. i Kosmonavtika (Rocket Technology and Astronautics)*, 14(2), 58–70.
35. Borisov, A. A., et al. (1981). On the regimes of fragmentation of drops and the criteria for their existence. *Inzhenerno-Fizicheskiy Zhurnal (Engineering and Physics Journal)*, 40(1), 64–70.
36. Gelfand, B. E., et al. (1971). Deformation of jets and drops of liquid in a drifting gas stream, Izvestiya AN SSSR. *Mekh. Zhidkosti i Gaza (Mechanics of Fluid and Gas)*, (3), 82–88.
37. Gelfand, B. E., et al. (1974). Varieties of crushing droplets in shock waves and their characteristics. *Inzhenerno-Fizicheskiy Zhurnal (Engineering and Physics Journal)*, 37(1), 119–126.
38. Ivandayev, A. I., et al. (1981). Gas dynamics of multiphase media. Shock and detonation waves in gas scales. In *Itogi nauki i tekhniki, VINITI, Mekh. zhidkosti i gaza* (The Results of Science and Technology, AISATI, Mechanics of Fluid and Gas, Moscow), 16, 209–290.

39. Korsunov, Y. A., & Tishin, A. P. (1974). An experimental study of the crushing of liquid droplets at low Reynolds numbers. *Izvestiya AN SSSR. Mekh. Zhidkosti i Gaza (Mechanics of Fluid and Gas)*, (2), 182–186.
40. (1982). *Itogi nauki i tekhniki, VINITI, Mekh. zhidkosti i gaza* (The Results of Science and Technology, AISATI, Mechanics of Fluid and Gas, Moscow), 17, 256.
41. Gelperin, N. I., et al. (1974). Spraying liquid with mechanical injectors. *Teor. Osnovy Khim. Tekhnol. (Theory Fundamentals of Chemical Technology)*, 8(3), 463–467.
42. Gelperin, N. I., et al. (1972). On the hydrodynamics of liquid-gas injectors with the dispersion of a working fluid. *Teor. Osnovy Khim. Tekhnol. (Theory Fundamentals of Chemical Technology)*, 6(3), 434–439.
43. Zvezdin, Y. G. (1972). *Investigation of a liquid-gas injector with dispersion of a working fluid.* Dissertation, Moscow Institute of Fine Chemical Technologies named after M.V. Lomonosov.
44. Zvezdin, Y. G., & Basargin, B. N. (1982). Hydrodynamic calculation of spraying of liquid by mechanical injectors. *Teor. Osnovy Khim. Tekhnol. (Theory Fundamentals of Chemical Technology)*, 16(5), 715–716.
45. Zvezdin, Y. G., et al. (1985). Hydrodynamics and heat exchange when spraying liquid in a high-temperature gas stream. *Teor. Osnovy Khim. Tekhnol. (Theory Fundamentals of Chemical Technology)*, 19(3), 354–359.
46. Basargin, B. N. (1974). *Investigation of hydrodynamics and mass-transfer capacity of injection-type devices.* Dissertation, Moscow Institute of Fine Chemical Technologies named after M.V. Lomonosov.
47. Rychkov, A. D., & Shraiber, A. A. (1985). Axisymmetric polydisperse two-phase flow with coagulation and fragmentation of particles for an arbitrary fragment distribution by mass and velocity. *Izvestiya AN SSSR. Mekh. Zhidkosti i Gaza (Mechanism of Fluid and Gas)*, (3), 73–79.
48. Sou, S. (1971). *Hydrodynamics of multiphase systems* (Russian Transl.). (Mir, Moscow).
49. Nigmatulin, R. I. (1978). *Fundamentals of mechanics of heterogeneous media.* Moscow: Nauka.
50. Belotserkovsky, O. M., & Davydov, Y. M. (1982). *The method of large particles in gas dynamics.* Moscow: Nauka.
51. Rivkin, V. Y., & Ryskin, G. M. (1976). Flow structure for the motion of a spherical droplet in a liquid medium in the region of transient Reynolds numbers. *Izvestiya AN SSSR. Mekh. Zhidkosti i Gaza (Mechanics of Fluid and Gas)*, (1), 9–15.
52. Ergun, S. (1952). Fluid flow through packed columns. *Chemical Engineering Progress*, 8(2), 89.
53. Abramovich, G. N. (1960). *The theory of turbulent jets.* Moscow: Fizmatgiz.
54. Abramovich, G. N., et al. (1975). *Turbulent currents under the influence of bulk forces and non-self-similarity.* Moscow: Mashinostroyeniye.
55. Abramovich, G. N. (1970). *On the influence of an admixture of solid particles or droplets on the structure of a turbulent gas jet.* DAN SSSR (Reports of AS USSR), 190(5), 1052–1055.
56. Abramovich, G. N., et al. (1972). Turbulent jet with heavy impurities. *Izvestiya AN SSSR. Mekh. Zhidkosti i Gaza (Mechanism of Fluid and Gas)*, (5), 41–49.
57. Abramovich, G. N., & Girshovich, T. A. (1972). The initial part of a turbulent jet containing heavy impurities in a spiral stream. In *Investigations of two-phase, magneto-hydrodynamic and swirling turbulent jets* (Proceedings of the MAI, Moscow), No. 40, pp. 5–24.
58. Abramovich, G. N., & Girshovich, T. A. (1973). *On the diffusion of heavy particles in turbulent flows.* DAN SSSR (Reports of AS USSR), 212(3), 573–576.
59. Aniskin, S. V. (1978). Similarity of the density of irrigation fluid sprayed by a mechanical injector SGP. In *Protection of the environment from pollution by industrial emissions in pulp and paper industry* (LTA, LTITSBP, Leningrad), No. 6, pp. 165–168.
60. Mikhailov, E. A., et al. (1981). *Development of a methodology for calculating the geometric dimensions of nozzles with a given character of the distribution of specific fluid flows* (Ruk. dep. ONITEKHIM, 20.04.1981, Yaroslavl), p. 6.

61. Mikhailov, E. A. (1982). *Research and development of a methodology for calculating apparatuses of chemical industries with a given character of the distribution of the density of irrigation.* Dissertation, Moscow Institute of Fine Chemical Technologies named after M.V. Lomonosov.

62. Abramovich, G. N. (1944). Theory of the centrifugal nozzle. In *Promyshennaya aerodinamika* (BIT TsAGI, Moscow).

63. Klyachko, L. S. (1952). The method of theoretical determination of the capacity of apparatus with a rotating axisymmetric flow of fluid. In *Theory and practice of dust-free ventilation* (Vol. 5, p. 162). Moscow: Profizdat.

64. Taylor, G. (1948). The mechanism of swirl atomizers. In *Proceedings of the 7th international congress for applied mechanics* (Vol. 2, pp. 280–285). London: The Congress.

65. Bammert, K. (1950). *Die Kern. Abmessungen in Kreisen den Stromungen,* Zeitschrift VDI, Bd. 92, No. 28, s. 32–39.

66. Rakhmatulin, H. A. (1956). Fundamentals of gas dynamics of interpenetrating motions of compressible media. *Jour. Prikladnaya Matematika i Mekhanika (Applied Mathematics and Mechanics), 20*(2), 184–185.

67. Krayko, A. N., et al. (1972). Mechanics of multiphase media. In *Itogi nauki i tekhniki, VINITI, Gidromekhanika* (The Results of Science and Technology, AISATI, Hydromechanics, Moscow), 6, 74.

68. Sternin, L. E., et al. (1980). *Two-phase mono- and polydisperse flows of gas with particles.* Moscow: Mashinostroyeniye.

69. Harlow, F. H. (1964). The particle-in-cell computing method for fluid dynamics. In *Methods in computational physics, vol. 3, Fundamental methods in hydrodynamics.* New York/London: Academic Press.

70. Amsden, A. A. (1966). *The particle-in-cell method for the calculation of the dynamics of compressible fluids.* Report LA-3466 (Los Alamos Science Lab, New Mexico).

71. Dukowicz, J. K. (1980). A particle-fluid numerical model for liquid sprays. *Journal of Computational Physics, 35*(2), 229–253.

72. Laats, M. K., & Frishman, F. A. (1970). On the assumptions used in the calculation of a two-phase jet. *Izvestiya AN SSSR. Mekh. Zhidkosti i Gaza (Mechanism of Fluid and Gas),* (2), 125–129.

73. Laats, M. K., & Frishman, F. A. (1973). Development of methods and investigation of intensity on the axis of a two-phase turbulent jet. *Izvestiya AN SSSR. Mekh. Zhidkosti i Gaza (Mechanism of Fluid and Gas),* (2), 153–157.

74. Abramovich, G. N. (1976). *Applied gas dynamics.* Moscow: Nauka.

75. Loitsyansky, G. G. (1978). *Mechanics of fluid and gas.* Moscow: Nauka.

76. Landau, L. D., & Lifshitz, E. M. (1953). *Continuum mechanics.* Moscow: Gostekhizdat.

77. Vlasov, V. V. (1975). *Investigation of the rectifying ability of the injector with dispersion of the liquid.* Dissertation, Moscow Institute of Fine Chemical Technologies named after M.V. Lomonosov.

78. Girba, E. A. (1978). *Investigation of the process of dust collection in liquid-gas injectors with dispersion of working fluid.* Dissertation, GIAP, Moscow.

79. Katalov, V. I. (1977). *Investigation of the absorption process in a liquid-gas injector with liquid dispersion.* Dissertation, MINH and GP im. I.M. Gubkin.

80. Basargin, B. N., & Katalov, V. I. (1975). Cell model of hydrodynamics of a free zone of a torch of injection devices. In *Macsoobmennye i teploobmennyye protsessy khim. tekhnol (Mass exchange and heat exchange processes of chemical technology)* (YaPI, Yaroslavl), pp. 65–74.

81. Leonchik, B. I., & Mayakin, V. P. (1981). *Measurements in disperse flows.* Moscow: Energoizdat.

82. Protodyakonov, I. O., & Glinsky, V. A. (1982). *Experimental studies of hydrodynamics of two-phase systems in engineering chemistry.* Leningrad: Leningrad State University.

83. Ametistov, E. V., et al. (1982). Heat and mass transfer. In V. A. Grigoriev & V. M. Zorin (Eds.), *Thermo technical experiment: Handbook.* Moscow: Energoizdat.

84. Fuks, N. A. (1975). Modern methods of aerosol research. *Zh. Vsesoyuzn. khim. obshchestva im. D.I. Mendeleyeva (Journal of Chemistry Society named afrter D.I. Mendeleyev), 20*(1), 71–76.

85. Basargin, B. N., et al. (1976). The velocity of droplets in a spray of a mechanical injector. In *Mass and heat exchange processes of chem. technol.* Interuniversity scientific research collection (YaPI, Yaroslavl), pp. 174–177.

86. Miller, M. (1979). *Holography.* (Transl. with czech). Leningrad: Mashinostroyeniye.

87. Rinkevichus, B. S. (1978). *Laser anemometry.* Moscow: Energia.

88. Malofeev, N. L., et al. (1981). The velocities of the motion of liquid droplets in a gas flow. *Zh. Prikl. Khim. (Journal of Applied Chemistry), 54*(2), 442–445.

89. Chigier, N. (1983). Drop size and velocity instrumentation. *Progress in Energy and Combustion Science, 9*(112), 155–177.

90. Zakharchenko, V. M. (1975). Measurement of the flow velocity by a laser one-beam time-of-flight method. Uchenyye zapiski TsAGI (Scientific Notes of CAHI), *6*(2), 147–157.

91. Zhigulev, S. V. (1982). On one version of the laser single-beam time-of-flight method for measuring the flow velocity. Uchenyye zapiski TsAGI (Scientific Notes of CAHI), *13*(5), 142–147.

92. Shifrin, K. S., & Golikov, V. I. (1961). Determination of the droplet spectrum by the method of small angles. In *Investigation of clouds, precipitation and thunderstorm electricity. Proceedings of the sixth interdepartmental conf.* (Izd. AN SSSR, Moscow), pp. 266–277.

93. Shifrin, K. S., & Kolmakov, I. B. (1967). Calculation of the particle size spectrum from the current and the integrand values of the indicatrix in the region of small angles. *Izvestiya AN SSSR. Fizika Atmosfery i Okeana (Physics of the Atmosphere and the Ocean), 3*(12), 1271–1279.

94. Bayvel, L. P., & Lagunov, A. S. (1977). *Measurement and control of the dispersion of particles by the light scattering method.* Moscow: Energia.

95. Dieck, R. H., & Roberts, R. L. (1970). The determination of the sauter mean droplet diameter in fuel nozzle sprays. *Applied Optics, 9*, 2007–2014.

96. Zimin, E. P., & Krugersky, A. M. (1977). Integral characteristics of light scattering by polydisperse particles. *Optika i Spektroskopiya (Optics and Spectroscopy), 43*(6), 1144–1149.

97. Battle, S. M., & Miller, T. J. (1981). Visualization of the transition region for flow around the wing profile with the help of smoke from a heated wire. *Missile Technology and Astronautics, 19*(4), 81–88.

98. Povkh, I. L. (1969). *Technical hydromechanics.* Leningrad: Mashinostroyeniye.

99. Rusanov, A. A., et al. (1969). *Ochistka dymovykh gazov v promyshlennoy energetike (Cleaning of flue gases in industrial power engineering).* Moscow: Energia.

100. Kremlevsky, P. P. (1980). *Measurement of flow and quantity of liquid, gas and steam.* Moscow: Izd-vo Standartov.

101. Leidenforst, W., & Ku, J. (1960). *New high-sensitivity micromanometer.* Instruments for scientific research (Russian Transl.). No. 10, pp. 76–78.

102. Ranz, W. E., & Marshall, W. R. (1952). Evaporation from drops (pt. 2). *Chemical Engineering Progress, 48*(5), 173–180.

103. Tikhonov, A. N., & Samarskii, A. A. (1977). *Equations of mathematical physics.* Moscow: Nauka.

104. Zueva, G. A. (2002). *Simulation of combined processes of heat treatment of heterogeneous systems intensified by combined energy supply.* Dissertation, IGXTU, Ivanovo.

105. Cranc, J. (1975). *The mathematics of diffusion* (2nd ed.). Oxford: Clarendon Press.

Chapter 2
Experimental Study of a Free Two-Phase Flow Generated by Spraying of Water in Air Using a Mechanical Injector

The basis of the method for calculating the processes of heat and/or mass transfer in a two-phase gas-droplet system consists of equations describing the hydrodynamics of such a system, taking into account the interfacial interaction. Knowledge and understanding of the mechanism of interaction of phases in a two-phase flow, which is created by spraying a liquid in a gas, is definitely not sufficient (as will be shown below). This is one of the main reasons for the abovementioned problem (see Chap. 1).

2.1 Measurements of Dispersion and Velocities of Droplets by Laser Methods

An experimental study of the hydrodynamics of a free two-phase flow produced by a mechanical nozzle was carried out in three stages. At the first stage, the dispersed characteristics of the atomized liquid were measured. At subsequent stages, the hydrodynamic characteristics of a dispersed liquid and gas were measured.

In the experiment, a two-phase stream produced by spraying of water in the air was studied at room temperature. To disperse the liquid, a centrifugal jet nozzle of *VTI* type was used (which was designed by the All-Russian Thermal Engineering Institute [1]) with an outlet hole diameter of 2 mm. In the experiment, the dispersion of the spray was measured: the size spectrum and the average Sauter diameter of droplets d_{32}. The radial profiles of the velocities, concentrations, and specific flows of the fluid and velocities and pressures of the gas were measured at different distances (up to 1 m) from the nozzle at various excess pressures $p = 300$, 500, and 900 kPa of a liquid on it, as well as changes in these magnitudes along the axis of the two-phase flow. Measurement errors generally did not exceed 5%.

All measurements in the spraying flow were carried out on the basis of the experimental setup shown in Fig. 2.1. Its main elements are the injector 1, the

© Springer Nature Switzerland AG 2020   31
N. N. Simakov, *Liquid Spray from Nozzles*, Innovation and Discovery in Russian Science and Engineering, https://doi.org/10.1007/978-3-030-12446-5_2

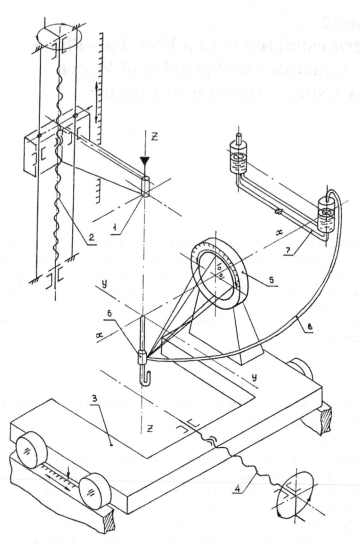

Fig. 2.1 The scheme of the experimental setup

drive 2 for vertical displacement of the nozzle allowing to change the axial coordinate z of the measure point, the platform 3 for positioning the measuring devices, and
the drive 4 for horizontal movement of the platform allowing to change the radial
coordinate r of the measuring point. Elements 5–8 relate to the measurement of the
hydrodynamic characteristics of the gas phase and will be described below.

To measure the dispersity of the atomizer spray, a method of small-angle light
scattering was used [2–6]. This method makes it possible to determine the spectrum
of the sizes of light scattering particles from the data of the scattering indicatrix of
laser radiation on droplets of a dispersed liquid.

Table 2.1 Mean diameter d_{32} of droplets in the spray flow in dependence on the pressure p of water in the nozzle and the distance z from it

p, kPa	300	500	900
z, mm	d_{32}, µm		
100	–	152	–
300	176	141	110
500	–	136	–
700	–	134	–

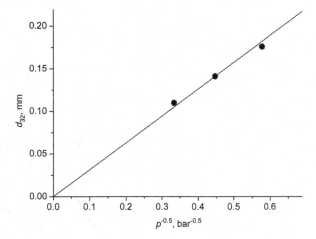

Fig. 2.2 Dependence of the average Sauter diameter d_{32} of droplets from the water pressure p in the nozzle at $z = 300$ mm

Experimental data for the average diameter d_{32} of the droplets (according to Sauter [5]) obtained in our experiments at different water pressures in the nozzle and at different distances from it are presented in Table 2.1 and in Fig. 2.2.

From Table 2.1 it is obvious that the droplet diameter d_{32} decreased with increasing liquid pressure in the nozzle and not significantly when moving away from it. The latter is explained by the secondary crushing of the droplets in the gas stream.

It is interesting to note that, according to Fig. 2.2, when the pressure of the liquid in the nozzle increases, the average droplet diameter d_{32} decreases approximately in inverse proportion to the square root of the pressure.

In the second part of the experiment to investigate the nozzle spray, the spatial distribution of the droplet velocities was studied. To measure the radial profiles of the axial component of the droplet velocity, a modification of the single-beam time-of-flight method was used [7, 8]. Its principal feature is as follows.

A laser emitting continuously at the main TEM00 mode creates a light beam with the Gaussian distribution of the radiation intensity along the radius of the beam

$$J = J_0 \times \exp\left[-(y^2 + z^2)/R^2\right] = J_0 \times \exp\left[-(r/R)^2\right]. \qquad (2.1)$$

where J and J_0 are the radiation intensities at some point of the beam and on its axis, respectively; $y, z, r = (y^2 + z^2)^{1/2}$ are the coordinates of points in the cross section of

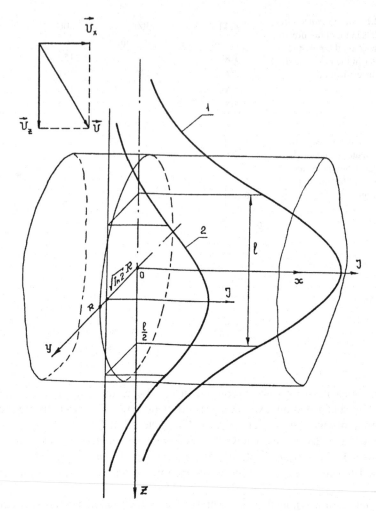

Fig. 2.3 The distribution of the light intensity J (y, z) in the cross section of the Gaussian laser beam, which determines the base distance l

the beam; and R is the characteristic radius of the beam. The meaning of R is that at a distance R from the beam axis, the radiation intensity J decreases by $e \approx 2.72$ times in comparison with the intensity J_0 on its axis.

The laser beam is directed so that in the measurement zone it is crossed by particles whose velocity is measured. We choose a rectangular coordinate system OXYZ with the center at the measurement point so that the OX axis coincides with the optical axis of the beam, and the velocity vector $\mathbf{U}$ of the particle is parallel to the OXZ plane (Fig. 2.3). In this case, the component U_z of the particle velocity is measured, and the component U_y is zero. Figure 2.3 shows the graphs of the function

$J(y, z)$ of the radiation intensity distribution in the cross section of a Gaussian beam. In the $y = \text{const}$ planes, curve 1 for $y = 0$ and curve 2 for $y = R \cdot (\ln 2)^{1/2}$ are located.

As the base we take the distance $l = z_1 - z_2$ between the points of the maximum change of the radiation intensity in the beam, i.e., points in which the absolute value of the first derivative along the direction OZ reaches a maximum and the second derivative equals to 0 and changes the sign when passing through these points. In accordance with this, we have

$$\partial J / \partial z = J \cdot \left(-2 \cdot z / R^2\right) \tag{2.2}$$

$$\partial^2 J / \partial z^2 = J \cdot \left(4 \cdot z^2 / R^4 - 2/R^2\right) = 0 \tag{2.3}$$

It follows from (2.3) that the value of the base distance $l = R \cdot 2^{1/2}$.

Obviously, the value l defined as the distance between the points with coordinates representing two solutions:

$$z_{1,2} = \pm R \cdot 2^{-1/2} = \pm l/2 \tag{2.4}$$

for Eq. (2.3) does not depend on the coordinate y of the intersection of the laser beam by the particle. And l is the distance given by Eq. (2.4) between two parallel planes horizontally dissecting the beam, which are equidistant from its axis.

The intensity I of the radiation scattered by the particle is proportional to the intensity of the light incident on it. Therefore, provided that the particle size is much smaller than the transverse dimension R of the laser beam, the variation in time of the radiation scattered by the moving particle is described by the same function as the change in the radiation intensity J in the cross section of laser beam, i.e., by the function of Gauss. Really

$$I = k \cdot J = k \cdot J_0 \cdot \exp\left(-y^2/R^2\right) \cdot \exp\left[-(U_z \cdot t)^2/R^2\right]$$

or

$$I = I_0 \cdot \exp\left(-t^2/T^2\right), \tag{2.5}$$

where k is the coefficient of light scattering, which depends on the properties of the particle and the radiation incident on it, U_z is the particle velocity component measured, $t = z/U_z$ is the particle flight time, $I_0 = k \cdot J_0 \cdot \exp(-y^2/R^2)$ is the maximum value of the radiation intensity scattered by the particle, and $T = R/U_z$ is the characteristic flight time of the particle.

In the same way as it was done in determining the basic distance, it is not difficult to establish from (2.5) the following. The duration τ of the light scattering pulse at the level of the maximum steepness of its leading and trailing fronts is equal to the time interval between two consecutive moments when the absolute values of the first derivative on time of the light scattering pulse $|dI/dt|$ are reaching the maximum value. In other words, the duration of the twice-differentiated light scattering pulse d^2I/dt^2 at the level of its zero value

$$\tau = T \cdot 2^{1/2} = l/U_z \qquad (2.6)$$

is nothing other than the time of flight by a particle of the basic distance l defined by formulas (2.3) and (2.4).

From (2.5) for $t_{1,2} = \pm\tau/2$, it is easy to establish that the intensity level of the radiation scattered by the particle, corresponding to the moments when the pulse of light scattering reach the maximum steepness of the fronts, is $I_0/e^{1/2} \cong 0.61 \cdot I_0$, i.e., 61% of the maximum equal to I_0. Thus, by measuring the duration of the light scattering pulse $I(t)$ at the level $I_0/e^{1/2}$, which is the same as the duration of the twice-differentiated pulse d^2I/dt^2 at zero level, it is possible to determine the time τ of the flight of a particle at the distance l. And by using the formula (2.6), one can calculate the value of the component measured U_z of the velocity of the particle crossing the laser beam.

Of all the light scattering signals on particles crossing the laser beam, single light pulse is optically extracted with a lens and a slit diaphragm from the measurement zone, the dimensions of which are limited by the diameter of the laser beam and the slit diaphragm width. The isolated pulsed light scattering signal on the particle is subjected to subsequent processing, which makes it possible to measure the duration of the light scattering pulse at the level $0.61 \cdot I_0$, i.e., at the level of the maximum steepness of its fronts. This is the main difference between the method used by us and the one described in [8], where the duration of the light scattering pulse was measured at the level $0.5 \cdot I_0$, i.e., at half of its amplitude.

This difference in the proposed and used by us modification of the laser single-beam time-of-flight method made it possible to process the pulsed light scattering signal, including double differentiation, in analog form with synchronous measurement of the duration of the received pulse by means of a secondary digital device. This allowed us to increase the speed of the measurements.

In Fig. 2.4 there is a schematic diagram of a device that provides the implementation of the above method for measuring droplet velocity in a spray of a nozzle.

The device consists of a laser 1, an objective 2, a slit diaphragm 3, a photodetector 4, an amplifier-converter 5, and a secondary digital device 6 for measuring the duration of electrical pulses. A Gaussian light beam of a He-Ne laser directed horizontally so that it intersects the axis of the nozzle and spray flow at a right angle at the required distance from the nozzle outlet creates a nonuniform in the vertical direction the light field defining the basic distance $l = R \cdot 2^{1/2}$. The droplets of sprayed water had dimensions of the order of 0.1 mm – considerably smaller than the diameter (~2 mm) of the laser beam. They moved almost rectilinearly in the "plane" formed by the light beam and the nozzle axis with a velocity of the order of 20 m/s, crossing the laser beam and scattering the radiation incident on them. The droplet velocity vector could be directed at an angle to the vertical in the range from zero to half of the angle $\theta = 65°$ near the root of the spray flow.

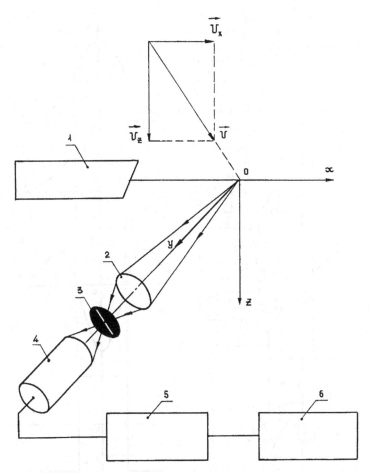

Fig. 2.4 Schematic diagram of the device for measuring of the particle velocity component U_z, which is perpendicular to the laser beam axis

The measurement zone, located at a distance of the order of 1 m from the laser, was separated optically by means of a lens 2 with a diameter of 45 mm with a focal length of 250 mm, tuned to a lateral scattering of light by particles under the angle of 90° to the beam axis, and a slit diaphragm 3 located in the plane of the image of the laser beam in the lens. The slit width was, approximately, 0.3 mm. The lens was located about 1 m from the laser beam. The "O" point of the intersection of the optical axis of the objective with the beam axis sets the center of the measurement zone. The edges of the diaphragm slit were oriented along the direction of flight of the particles and determined the size – about 1 mm – of the measurement zone along the laser beam axis.

The pulsed light scattering signal extracted with an objective and the slit diaphragm on particles passing through the reference distance in the measurement zone was transformed by a photomultiplier 4 into electric current pulses having an average

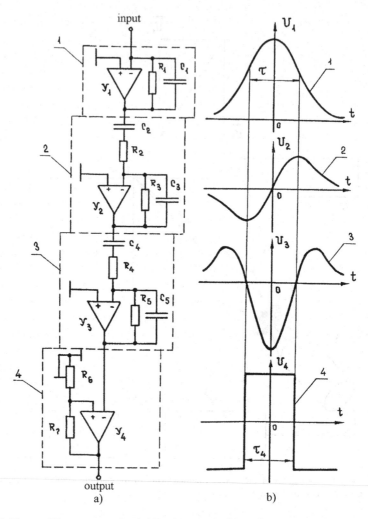

Fig. 2.5 The amplifier-converter circuit (**a**) and the shape of the voltage pulses at the output of each cascade (**b**)

amplitude of 0.5 μA and a Gaussian shape (5) as the light scattering pulses. The signal from the output of the photodetector 4 was subjected to further processing in a specially designed for these measurements and manufactured amplifier-converter 5, which formed rectangular pulses of the same amplitude of 10 V, the durations of which of the order of 100 μs, uniquely related to the time-of-flight by the particles of the base distance, was measured by a digital frequency meter 6.

The simplified electrical circuit of the amplifier-converter, which consists of four series connected cascades 1–4 made on the basis of operational amplifiers Y_1–Y_4, is shown in Fig. 2.5a.

Figure 2.5b illustrates a successive change in the shape of the pulse signal in the amplifier-converter: each of the curves 1–4 shows the pulse shape at the output of the corresponding cascade.

The first cascade, which is a low-frequency amplifier built on a Y_1 chip, coordinated the high impedance output of the photomultiplier with the low-impedance input of the next cascade. Identical to each other, the second and third cascades, made on the basis of operational amplifiers Y_2 and Y_3, twice sequentially differentiated the electric pulse passing through them.

The differentiating properties of these cascades are due to the linear increase in their amplitude-frequency characteristic in the low-frequency region, which is the main spectral region of the pulse signal of the Gaussian form. This type of amplitude-frequency characteristic of cascades 2 and 3 was provided by capacitive character (capacitors $C_2 = C_4 = 47$ nF) of the input impedance at low frequencies and the active impedance (resistance $R_3 = R_5 = 5.6$ kΩ) of the negative feedback circuit. The introduction of capacitors $C_3 = C_5 = 2200$ pF and resistances $R_2 = R_4 = 270$ Ω into these cascades, as well as the inclusion of the capacitor $C_1 = 112$ pF in the feedback circuit of the first matching cascade, served to decrease the values of the amplitude-frequency characteristics of these cascades at high frequency and thus for suppression of high-frequency noise of the signal.

The impulse signal from the output of the second differentiating cascade (3) enters the input of the phase detector (4), represented in this case by a Schmitt trigger, built on the basis of the electronic microcircuit Y_4. The resistor ratings of the voltage divider in the positive feedback circuit are as follows: $R_6 = 4.7$ kΩ and $R_7 = 1.5$ MΩ. At the output of the last cascade of the amplifier-converter, rectangular pulses of the same amplitude were obtained with duration equal to the time interval between two successive transitions through the zero level of the impulse voltage signal at the output of the second differentiating cascade 3. The variable resistance R_6 made it possible to adjust the trigger threshold of the Schmitt trigger and resolution of the amplifier-converter over the amplitude of pulses of the light scattering signal. This provided the possibility of regulating the overall resolution of the device over the number of pulses of light scattering emitted by the particles that are simultaneously in the measuring zone by discrimination of all pulses whose amplitude at the input of the Schmitt trigger was below the threshold of its triggering.

It should be noted that between the duration τ_4 of the rectangular pulses generated by the amplifier and the duration τ of the pulses of light scattering at the level of the maximum steepness of the fronts, the exact equality (independent of the numerical value of the time τ of flight by the particle through the base distance) could occur only under the condition of ideal linearity of the pulses transformation preceding the phase detecting in the last cascade 4 of the amplifier-converter. The high speed of the photoelectric multiplier provided with sufficient accuracy a linear photoelectric conversion of the light scattering pulse with a low-frequency spectrum. However, it was impossible to ensure sufficient linearity of the signal conversion in the amplifying 1 and, especially, in differentiating cascades 2 and 3 of the amplifier-converter, due to the necessity to provide the decrease the values of amplitude-frequency characteristics of these cascades at high frequency in order to suppress high-frequency noise and eliminate self-excitation of the amplifier-converter. For this reason, the values of the duration's τ and τ_4 coincided only for one numerical value of τ, and for all others, they differed slightly in one direction or another. At the same time, the parameters of the amplifier-converter circuit were calculated in such a way that the values of τ and τ_4 were in one-to-one correspondence $\tau = f(\tau_4)$ in a sufficiently wide interval of values of the time-of-flight by particles through the base distance.

In accordance with (2.6), in order to determine the velocity U_z of particle by using the time τ of its flight through the base distance l by the formula

$$U_z = l/\tau = l/f(\tau_4) \tag{2.7}$$

it is necessary to know the value and the function $f(\tau_4)$ connecting the quantities τ and τ_4. For measuring l, it is possible to calibrate the laser beam by photometry, for example, by means of a photomultiplier with a "point" diaphragm moving in the measurement region across the beam. Having thus obtained the distribution (2.1), one can determine from it R and $l = R \cdot 2^{1/2}$.

With respect to the dependence $\tau = f(\tau_4)$, the matter is more complicated, since for its determination it is necessary to calibrate the duration τ of the input pulses versus a duration of τ_4 output pulses of the amplifier-converter when a signal is fed to its input from the generator of pulses of the Gaussian shape and controlled duration. But such generators currently do not exist. For this reason, the whole device was calibrated by means of the function

$$U_z = F(\tau_4), \tag{2.8}$$

which has connected the velocity U_z of intersection by particle a laser beam with a duration of τ_4 rectangular pulses formed by an amplifier-converter from light scattering signals.

The advantage of such calibration is that it automatically takes into account the nonlinear distortions in the processing of light scattering pulses.

The calibration was carried out using a thin opaque disk with a 0.2 mm hole at a distance ρ from the center of the disk. When a disk rotated with some frequency ω in a plane perpendicular to the beam axis, the hole, crossing the beam across, modeled the particle passing at velocity $U_z = \omega \cdot \rho$. The radiation passed through the hole, varying in time in accordance with the same function (5), as that scattered on the particles, was perceived by a photomultiplier. The impulse electric signal of the photomultiplier was subjected to the same treatment as in the case of the pulsed light scattering signal on the particles. The duration of τ_4 pulses at the output of the amplifier-converter, as well as frequency $n = \omega/2\pi$ of the pulses of light radiation passing through the hole in the disk, was measured by a digital frequency meter. The radius $\rho = 150$ mm of a circumference described by the hole was much larger than the transverse dimension $R = 1$ mm of the laser beam, which ensured a practically straight line intersection of the beam with the hole. The disk rotation frequency n was

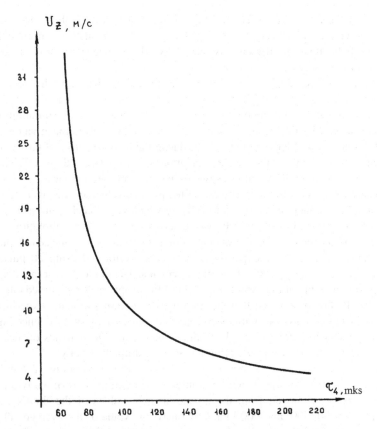

Fig. 2.6 Calibration curve of the device for measuring the velocity of particles using their time-of-flight through single laser beam

varied by a reducer in the range from 3 to 30 s^{-1}, which corresponded to a change in the hole speed from 3 to 32 m/s.

The obtained by the described method calibration curve of the device for measuring the particle velocity component transverse to the laser beam is shown in Fig. 2.6.

The abscissa shows the values of the duration τ_4 of rectangular pulses at the output of the amplifier-converter and along the ordinate – the one-to-one corresponding to them the values of the velocity U_z of the hole in the disk. When measuring the droplet velocity in the spray of the nozzle, the speed values were determined using the calibration curve of the device from the value of duration τ_4 of the rectangular pulses, formed by the amplifier-converter and measured by the frequency meter.

The above method for measuring the particle velocity component U_z perpendicular to the axis of the laser beam was used to study the distribution of the axial velocity component of the disperse phase in the spray of a mechanical nozzle.

Changes of the droplet velocity on the vertical axis of the two-phase flow versus height, as well as radial changes of the velocity axial component in different cross sections of the flow at different distances from the nozzle, were investigated at several fluid pressures on the nozzle.

The study showed that at each point of the spray flow, the liquid droplets move with very different speeds. This is evidenced by the significant difference in the duration of the light scattering pulses: the standard deviation σ_τ of the durations of τ_4 pulses, measured by the frequency meter, was about 25% of the mean value $<\tau_4>$. This difference in droplet velocities can be due to the random nature of the liquid jet disintegration, as well as the pressure fluctuations on the nozzle and the difference in the drag of droplets of different sizes when they interact with the gas.

Because of the statistical distribution of the particle velocities, the average value of the axial component of the particle velocity was taken at each point as the final result of the measurement, which was determined using the calibration curve (Fig. 2.6) of the device for the average duration $<\tau_4>$ of the measured pulses. Statistical averaging of the experimental data was performed using 20 pulses of light scattering at each "point" of the measurements. The random error in determining the average value of the axial component of the droplet velocity turned out to be about 5%. It should be noted that the used method permit automatic experimental data processing synchronous with measurement of pulses duration by connecting the frequency meter to a computer, which allows increasing the statistics of measurements to reduce the error in measuring the average droplet velocity.

The results of the studies are presented in the form of graphs in Figs. 2.7 and 2.8.

Figure 2.7 illustrates the change in droplet velocity on the axis of the two-phase flow as they are removed from the nozzle at three values of the fluid's excess pressure $p = 300$, 500, and 900 kPa. Obviously, the value of the average velocity of drops decreases in the coordinate axis z as a result of their interaction with air. The relative decrease in the velocity of drops on the axis of the flow on a 100–900 mm interval from the nozzle is small, about 20%.

Figure 2.8 shows the profiles of the radial change in the axial component of the droplet velocity measured at a liquid pressure at the nozzle $p = 500$ kPa. The straight line inclined to the axis OZ in Fig. 2.8 shows the conditional boundary of the two-phase flow, determined by the surface of the cone with the opening angle, equal to the measured root angle of the spray $\varphi = 65°$. The shape of the velocity profiles taken at different distances $z = 100$, 300, 500, and 700 mm from the nozzle is obviously the same and indicates a decrease in the measured component of the droplet velocity over the radius of the flow.

The decrease in the axial component of the droplet velocity over the radius of the flow, and the fact that the profiles have a concave shape ($\partial^2 U_z/\partial r^2 > 0$), can be explained by a similar kind of the initial velocity distribution of the fluid over the radius of the nozzle outlet hole and, in part, by the differences in drag of droplet at their interaction with gas. The deceleration of droplets is greater at the periphery of the flow, where, according to experimental data, the gas velocity is lower than in the near-axis zone (see below, Sect. 2.2).

Fig. 2.7 Variation of the
droplet velocity along
the two-phase flow axis
at the nozzle pressure
p: ○ – 300, △ – 500,
□ – 900 kPa [9]

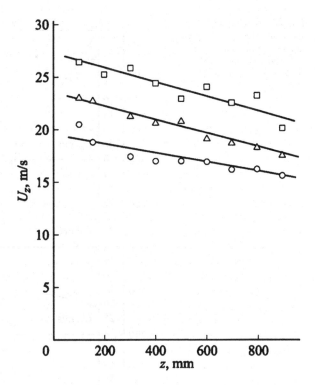

The following conclusions can be drawn, based on the results of measurements, presented in Sect. 2.1.

1. The velocities of the droplets dispersed by a nozzle at each point of the flow are distributed in a rather wide range $\sigma_U \cong 0.25 <U_z>$.
2. The average velocity of droplets decreases over the axis of the flow as a result of their interaction with gas. The relative decrease in speed is small: for droplets moving along the axis of the nozzle, it constitutes about 20% in the investigated range.
3. The average value of the axial component of droplet velocity also decreases (by about 30%) along the radius of the flow in the direction from its axis to the periphery. This is mainly explained by the initial velocity distribution of the liquid that was present at the outlet from the nozzle and partly due to an increase with the distance from a nozzle of the droplet deceleration effect when they interact with the gas.

The results presented in this section were previously published partially in articles [9, 10].

Fig. 2.8 Radial profiles of the axial velocity of droplets at a nozzle pressure $p = 500$ kPa, line 1 – the boundary of the flow of droplets [9]

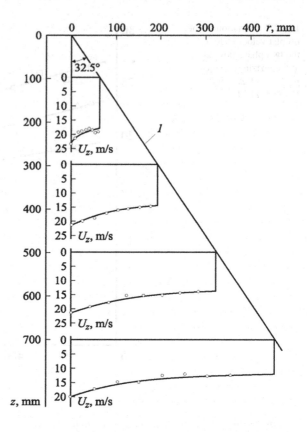

2.2 Measurements of Hydrodynamic Characteristics of the Gas Phase by Pneumometric Methods

The purpose of the third stage of experimental studies of the structure of a two-phase flow in a spray of a nozzle was the study of velocity fields and static gas pressures. To do this, it was necessary to improve the devices for using the pneumometric method of measuring the gas flow in the spraying, including receivers of full and static gas pressures and a capillary micromanometer.

The unity of experimental methods for measuring velocity fields and static gas pressures in the spray stream was provided by using the pneumometric Pitot-Prandtl tubes in combination with a precision micromanometer.

It was found that tubes of known types [11] cannot be used for measurements in two-phase disperse flows with a relatively high concentration of a dispersed liquid phase. In particular, when measuring in the spray of the nozzle, it was found that the internal full-pressure cavity of the known tubes was filled with liquid and the static pressure-receiving holes were covered by a liquid film flowing along the outer surface of the tubes; this disrupted the operation of the pressure receivers. In addition, it was found that the ingress of liquid droplets through the intake opening into the full-pressure cavity of the pneumometric tube and their braking as a result of interaction with the gas located there can lead to an additional compression of the gas in the total pressure cavity and thereby to an overestimation of the manometer reading when measuring the gas velocity.

In order to eliminate these drawbacks and enable the measurement of velocities and static gas pressures in a two-phase flow, improvements were made to the known pneumometric tubes and a micromanometer, which were as follows.

First, in order to increase the transverse dimensions of the internal cavities and thereby prevent the formation of liquid "plugs" in them, which, due to capillary phenomena, introduce an error in the pressure measurement, the receivers of the total and static pressure were made in the form of two separate tubes. In this case, the ratio of the characteristic constructive dimensions – the length of the head, the outer diameter, the diameters of the receiving openings of the total and static pressures, and the distance from the front end of the tube to the receiving openings of the static pressure – was chosen in accordance with a standard tube with a hemispherical tip [11]. The outer diameter of both tubes was 10 mm. The design of full and static pressure receivers, allowing measurements in a gas flow containing droplets of liquid, is shown in Fig. 2.9.

Second, in order to prevent the liquid filling of the internal cavities of the receivers of the full and static pressures, the lower part of each head 1 and 2 at a distance of 15 outer diameters from the receiving holes was provided with an additional capacity 3 equipped with a drain pipe 5 attached to it by means of a flexible hose 6 and a hydraulic seal 7. The measuring connections 4 connecting the internal cavity of the pressure receiver with the micromanometer were built into this additional capacity. Moreover, their inner ends were placed inside the cavity, and the inlet holes were made with a sloped end face in the direction opposite to the liquid flow, which prevented moisture from entering the measuring line.

For the convenience of sequential measurements of the total and static pressures, the additional capacitance 3 was made up of two separable parts. The upper part in the form of a cylinder interfaced with a cone was connected to the lower end of each of the heads. The lower part in the form of a cylinder interfaced with the hemisphere was equipped with a hydraulic seal and a measuring branch pipe and fixed to a special rotating device mounted in the corresponding place of the flow (Fig. 2.1). The threaded connection of the upper and lower parts of the additional capacitance ensured the interchangeability of the receivers of the total and static pressure when measured at the same point of flow.

The fact that the standard pressure tube was adopted as a basis for improved pressure receivers [11] made it possible to avoid the need for special calibration for the pressure perceived and transmitted to the secondary device.

Experimental study of the effect of additional gas compression by braking in the cavity of full pressure the droplets of dispersed liquid showed that a full-pressure tube with an internal cylindrical channel diameter equal to the diameter of the hole receiving the total pressure gives the increased pressure of up to 15% at a distance of

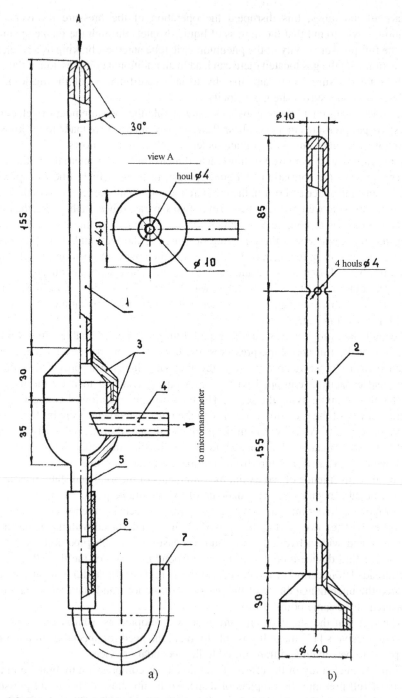

Fig. 2.9 Receivers of total (**a**) and static (**b**) gas pressure in a two-phase disperse flow

150 mm from the nozzle in comparison with the tubes in the construction of which measures were taken to reduce this effect. To reduce this, it was necessary to expand the internal volume of the total pressure receiver by increasing the diameter of its cylindrical channel with respect to the diameter of the receiving hole. An additional measure was that the edge of the intake opening of the total pressure was made sharp. Their shape was determined by the intersection of the outer hemisphere with a diameter of 10 mm and an inner cone with an opening angle of 60° (Fig. 2.9a).

In the initially selected design of the static pressure receiver, the head diameter was 10 mm, and eight static pressure-receiving holes had a diameter of 1.3 mm, the value of which was related by the standard ratio [11] with the head diameter. The experimental verification of such a receiver was made in a stream of air containing water droplets and shows that the receiving holes were overlapped by a liquid film, which violated the operability of the receiver. It was not possible to eliminate this phenomenon by selecting the material to cover the external surface of the static pressure receiver. For this aim some hydrophobic materials were tested: fluoroplastic, paraffin, and polyethylene wax, and as hydrophilic substance - kerosene. Therefore, the number and diameter of the receiving openings of the pneumometric tube for static pressure had to be made different from the standard values [11]. Thus, a decrease in the number of symmetrically located holes from 8 to 4 and an increase in their diameter from 1.3 to 4 mm with a decrease in the thickness of the wall of the receiver body at the location of the holes from 1 to 0.3 mm prevented the formation of films that overlap the receiving apertures of the static pressure.

In order to elucidate the nature and degree of influence of the constructive changes in the static pressure receiver on its operation, we carried out an experimental comparison of standard and modified structure tubes under the same conditions. The tubes were tested in a stream of air, which blown through the pipe and did not contain liquid droplets. As a result, it was found that the readings of the receiver of the modified design were 4–15% (on average 10%) lower than the standard receiver. This means that a modified receiver made it possible to measure static pressure in a stream containing drops of liquid with satisfactory accuracy even without additional calibration. To obtain more accurate experimental data, the calibration curve obtained in the tests was used, and the correction was made to the result of the measurement by a pneumometric tube of a changed design.

An important constructive feature of the Pitot-Prandtl velocity tube [1, 11] is the placement of the static pressure-receiving holes at a sufficiently large distance h downstream from the full-pressure-receiving hole located on the end surface of the tube. This is due to the need to establish a flow and equalize the velocities in it after flowing around the end surface of the tube before the stream reaches the receiving apertures of the static pressure. It is clear that this circumstance plays a less important role in the case of a homogeneous gas flow, for example, in a pipeline of constant cross section which diameter is larger than the dimensions of the Pitot-Prandtl tube.

For the value of the distance h from the end of the tube to the receiving apertures of static pressure, different values are recommended in the literature. So, in [1, 12] it was recommended h to be equal to 3 tube diameters and in [11] for tubes with different shapes tip – conical, spherical, and ellipsoidal – no less than 6–8 tube diameters.

The author of this work carried out a theoretical analysis of the effect of this distance h on the error in measuring the static pressure in a gas flow on the basis of a consideration of the potential flow near a semi-infinite body. As a result, it was found out that for measuring the gas velocity with an error of not more than 1%, it is sufficient to have a distance h between the holes of full and static pressure equal to 3 tube diameters. In this case, the error in measuring the static pressure with such a tube is proportional to the dynamic pressure $P_D = \rho \cdot W^2/2$ and decreases with increasing distance h.

To reduce the relative errors in the measurement of the degree of the gas rarefaction ΔP in the two-phase flow, proportional to the ratio of $P_D/\Delta P$ that can be significant, it is necessary to use a pressure receiver with a large distance h between the static pressure-receiving holes and the tube end. So, at liquid pressure $p = 500$ kPa in the nozzle at the distance $z = 150$ mm from it on the axis of the two-phase flow the ratio $P_D/\Delta P$ was approximately equal to 35. The relative error in measuring the rarefaction ΔP by a tube with receiving apertures remote by 3 diameters from the end would be 50%, and for a tube whose holes are removed by 9 diameters the error is less than 6%. Therefore, in the experiment, the tube was basically used, which is closer to the recommended standard [11] with a distance h between receiving holes and a streamlined end equal to 8.5 tube diameters. In this case, the point at which the symmetry axis of these holes intersected was considered as the measurement point; this is more natural than the point of location of the tube end, although not completely unique in the inhomogeneous divergent gas flow. With this determination of the measurement point, it was assumed that the alignment of the flow parameters after flowing around the tube end occurs simultaneously with their variation due to the expansion of the flow and is completed by the time it reaches the receiving holes of the static pressure.

With a tube diameter of 10 mm, its end protruding in front of the receiving apertures had a length $h = 85$ mm, which made impossible to use this tube for the measurements of the air pressure near the root of the spray due to a significant perturbation of the flow when the droplets of dispersed liquid were collided with the surface of the receiver. For this reason, using this tube, the static pressure measurements were carried out only at points of the two-phase flow far enough from the nozzle, at distances $z \geq 150$ mm, and the results of these measurements presented below can be considered as quite correct.

When investigating the nature of the air entrance from the surrounding space to the inside volume of two-phase flow and air motion near the spray boundary, due to a significant decrease in the speed and static air pressure when moving away from the nozzle, the sensitivity of the commonly used micromanometer of the MMN type was insufficient for measuring at distances $z \geq 150$ mm from the nozzle. But it allowed performing them in a region closer to the nozzle. To carry out this investigation, with a moderate perturbation of the flow by the pressure receiver, another static pressure tube was made. It differed from the tube used in the rest of the measurements and presented in Fig. 2.9b only that its end protruding in front of the static apertures was shorter. Its length h was not 85 mm, but 30 mm or 3 tube diameters, as in the high-speed Pitot-Prandtl tubes described in [1, 12].

The drawback of this tube, which consists in a greater error in the measurement of the static pressure, was partially compensated for by its smaller perturbing effect on the flow. Nevertheless, the profile of the static gas pressure measured at the distance $z = 100$ mm from the nozzle, measured with this truncated tube, and presented below together with other results, cannot be considered quite correct and should be regarded as a qualitative dependence.

A comparison of both the tubes used for measuring static pressure, the main one (Fig. 2.9b) and the shortened one, was carried out at the same points of the spray nozzle at distances $z = 150$ mm from the nozzle. It showed that the direction of the flow of both tubes is perceived identically, and the values of the pressure fixed by the micromanometer can differ by up to $\pm 40\%$. In comparison with the main tube, the truncated one with significant rarefaction and high flow velocities in relative proximity to the injector gives underestimated readings and at low pressures and gas velocities at more distant flow points – understated readings. The reason for this difference in the measuring results, besides the difference in the length of the tubes, is probably also the expansion of the gas flow in the divergent spray torch.

As noted above, the error in measuring the gas velocity is less dependent on the distance between the receiving static pressure holes and the end of the Pitot-Prandtl tube than the measurement error of the static pressure. This circumstance made it possible to consider the experimental data obtained by the combined use of a shortened tube of static pressure and a full-pressure tube (Fig. 2.9a), sufficiently correct to determine the gas velocity in the cross section of the two-phase flow at a distance $z = 100$ mm from the nozzle and also near the conditional boundary of the stream.

These results, together with others relating to the determination of the gas velocity in the regions of the flow which are more farther from the nozzle using the main tube for measuring the static pressure and the tube of full pressure (Fig. 2.9), are given below.

In addition to the above-described improvements in the receivers of the total and static gas pressures, it was necessary to perform the improvement of the differential micromanometer to measure these quantities. The modification of a highly sensitive capillary micromanometer used in our experiment is described below.

The degree of the gas rarefaction in the internal volume of the free two-phase flow produced by the nozzle with respect to the gas in the surrounding space is relatively small. Taking into account the Bernoulli equation for the gas inflowing into spray stream, it is natural to assume that the pressure difference between the peripheral and the near-axis zones of the flow, which characterizes the degree of rarefaction of the gas in the flow, is of the same order of magnitude as the value $P_D = \rho \cdot W^2/2$ near the flow axis. At liquid exit velocities of about 30 m/s (which corresponds to an excess pressure in the nozzle of the order of several atmospheres), the velocity of the gas in the near-axis zone near the root of the spray flow will be somewhat lower but of the same order as the velocity of liquid flowing from the nozzle. In this case, the gas

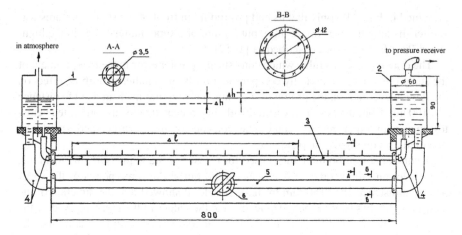

Fig. 2.10 Highly sensitive capillary micromanometer

velocity in the peripheral regions of the two-phase flow can be less than 5 m/s, which is confirmed by experiment [13]. At such gas velocities, its speed pressure and the degree of rarefaction do not exceed 20 Pa. At the same time, the industrially manufactured devices at our disposal did not allow us to measure such small pressure differences with sufficient accuracy. For example, an inclined micromanometer provided an error of not more than 5% when measuring the pressure drop of at least 20 Pa or 2 mm of water column.

To ensure the accuracy and reliability of measurements of lower pressures, a highly sensitive capillary micromanometer was made, whose design is fundamentally analogous to the device described in [14]. The device of the micromanometer is shown in Fig. 2.10.

The device consists of two identical cylindrical vessels 1 and 2, which are connected by means of a thin glass capillary 3. The capillary is connected to the vessels by means of flexible rubber hoses 4. In addition to the capillary, the vessels are connected using the tube 5 equipped with a tap 6 and having a cross section of the internal channel substantially greater than the cross section of the capillary. The tube 5 performs an auxiliary function: it makes it possible, by opening the tap 6, to achieve a rapid equilibrium position of the fluid levels in the vessels.

The principle of the device is as follows.

Using a syringe through the rubber tube 4, a bubble of air is introduced into the capillary, the initial position of which is set and fixed using a tap 6. At a bubble length that is about ten times greater than its diameter equal to the diameter of the capillary and sufficiently low velocity of liquid flow in the capillary, the air bubble acts as a piston without leakage. When a pressure difference $\Delta P = P_1 - P_2 > 0$ is created above the liquid surfaces in the vessels 1 and 2, the liquid level in them will, respectively, drop or rise by the value $\Delta h = \Delta P/(2 \cdot \rho_1 \cdot g)$, where ρ_1 is the density of fluid in the vessels and g is the acceleration of gravity. The air bubble then moves inside the capillary, and during the establishment of equilibrium in the system, it moves to a distance $\Delta l = \Delta h \cdot D^2/d^2$, where D and d are the diameters of the vessels and capillary, respectively. The cross-sectional ratio of the vessels and the capillary D^2/d^2 can be chosen sufficiently large, which makes it possible in the device of this design to achieve a very high sensitivity. Thus, in the device described in [14], the ratio D^2/d^2, representing the coefficient of increase in the change in liquid levels Δh, was of the order of 10^4, so that the device made it possible to measure pressure drops with an absolutely error of approximately 10^{-4} mm of water column.

However, possessing high sensitivity, the device mentioned, apparently, was too inertial. Considering the fluid motion in the capillary as quasi-stationary, and using the Poiseuille formula for the viscous friction force of a fluid on the capillary wall, one can obtain from the fluid motion equation an expression for the relaxation time in a system of capillary-connected vessels

$$\tau = \frac{16 \cdot \mu \cdot L \cdot D^2}{\rho_{\text{ж}} \cdot g \cdot d^4} \tag{2.9}$$

where μ is the viscosity of the liquid and L is the length of the capillary.

In the device described in [14], ethyl alcohol was used as the working liquid, in which the density $\rho_1 = 790$ kg/m^3 and viscosity $\mu = 1.22 \cdot 10^{-3}$ Pa·s. The design of the instrument had such parameters: capillary length $L = 5$ m, ratio of the cross sections of vessels and capillary $D^2/d^2 = 10^4$, and capillary diameter $d = 2$ mm. From these data, taking into account the value $g = 9.8$ m/s^2, the estimate for the relaxation time of the device gives $\tau = 3 \cdot 10^4$ s, i.e., about 8 h.

In order to increase the speed of measuring, the design dimensions of the instrument were changed to the following: the diameter of the vessels $D = 59$ mm, the diameter of the capillary $d = 3.4$ mm, the ratio $D^2/d^2 = 300$, the length of the capillary $L = 840$ mm, and the diameter of the tube 5 for the rapid establishment of the same level of fluid in the vessels was 10 mm. In addition, as working liquid, acetone was used, which is comparable in density and surface tension to ethanol used in the instrument [14], but has a significantly lower viscosity $\mu = 3.3 \cdot 10^{-4}$ Pa·s.

The introduction of these changes in the design of a precision micromanometer enabled to decrease the relaxation time τ to 15 s and made it possible to measure pressure drops in the gas with an error of no more than 5% in the range of 0.7–40 Pa or corresponding air velocities in the range from 0.9 to 8 m/s with an error of no more than 2.5%. The joint use of an industrial device – an inclined micromanometer type MMN – and 40 times more sensitive instrument, the above-described capillary micromanometer, made it possible to measure gas velocities in a wide range of 0.9–55 m/s with an error not exceeding 2.5%. The technique of pneumometric measurements is described below.

To study the fields of velocities and gas pressure in two-phase flow produced by the nozzle and for measuring characteristics of dispersed phase of the stream, the same installation was used (Fig. 2.1). In measuring the hydrodynamic characteristics of the gas phase, it additionally included a special rotation device 5 which makes it possible to vary the angle of inclination of the pneumometric tube 6 to the axis of the flow and a micromanometer 7 connected to the pneumometric tube by means of a flexible hose 8. The rotation device 5 consisted of a ball bearing, which was rigidly fixed in the support, and on the inner ring was mounted a rigid holder for the pressure receiver. The angle of inclination of the pressure tube changed when it was rotated around the bearing axis XX, passing through the receiving holes, and was measured by the measuring scale.

In the course of the experiment, the values of the static and total air pressure at various points of the two-phase jet were measured by means of appropriate pressure receivers and a micromanometer at three values of the excess water pressure in the nozzle $p = 300, 500,$ and 900 kPa. In this case, the pressure perceived by the receiver at the measuring point was compared by a micromanometer with the air pressure above the surface of the liquid in the open vessel of the micromanometer, which was located 2 m from the axis of the torch. The axial coordinate z of the measuring point with respect to the nozzle was changed by the vertical displacement of the latter relative to the pneumometric tube fixed in this direction. The initial position of the air bubble in the capillary was established and fixed with the injector turned off. And the readout of total excess air pressure ΔP_{full} (or its rarefaction ΔP) sensed by the receiver at the measuring point when the nozzle already operated was made with respect to this "zero" (initial) reading of the micromanometer. Thus, the micromanometer was measuring the change in gas pressure that occurs in the cavity of the receiver located at the measurement point after the injector is switched on.

The described technique realizes the measurement of rarefaction, caused solely by the interaction of liquid and gas inside of the two-phase jet. If necessary, it is also possible to take into account the hydrostatic component of pressure ΔP_z relative to the surface of any level, for example, relative to the plane of the nozzle outlet, equal to $\Delta P_z = \rho_g \cdot g \cdot z$, where z is the distance from the measuring point to the plane of the nozzle outlet.

As a measuring result of the gas static pressure ΔP, the maximum rarefaction reading was recorded corresponding to the position of the pneumometric tube, in which the latter is oriented in the direction of air movement. When working with the full-pressure receiver, the maximum reading of the micromanometer, corresponding to the orientation of the pneumometric tube along the air current lines, was also taken as the measurement result ΔP_{full}.

It should be noted that the sensitivity of the readings of the micromanometer to the orientation of the pneumometric tube in the gas flow was directly dependent on the magnitude of the pressure it perceived and the sensitivity of the micromanometer used. So, on the flow axis near the nozzle, where, due to relatively large full-pressure values, the measurement was conducted with an inclined micromanometer of the MMN type, the sensitivity of its readings to the orientation of the full-pressure tube was significantly lower than that of the capillary micromanometer. The latter was used in conjunction with the corresponding pneumometric tube for measurements of the static gas pressure at the same flow points. In the rest of the spray area, where the values of the total pressure did not exceed the range of the capillary micromanometer, this device was used to measure both the total and static gas pressures. The sensitivity of its readings to the change in the orientation of the full-pressure pneumometric tube was higher than for the static pressure tube, in accordance with the difference in the magnitude of the measured pressure drops. At the same time, at each of the measurement points, the angle of inclination of both pneumometric tubes to the axis of the flare, which determines the direction of the gas velocity vector, turned out to be the same within the measurement errors ($\pm 0.5^\circ$) or the setting (up to $\pm 2^\circ$) of this angle at the maximum of the readings of the micromanometer.

The air velocity was calculated from the values obtained by measuring its total pressure ΔP_{full} and the static ΔP pressure in accordance with equation:

$$W = \sqrt{\frac{2 \cdot (\Delta P_{\text{full}} + \Delta P)}{\rho_{\text{g}}}} \qquad (2.10)$$

The sign "+" in formula (2.10) considers the fact that the modulus of the rarefaction of the gas in the flow is denoted by ΔP.

The results of the experiment on the measurement of the velocity and gas pressure fields in the spray of the nozzle, carried out as described above, are presented in the form of graphs in Figs. 2.11, 2.12, 2.13, 2.14, and 2.15.

Measurements of the static gas pressure in the sprayed liquid jet showed that as a result of the interaction of the dispersed liquid and gas in the flow a rarefaction of gas is created, the value of which depends on the pressure of the liquid in the nozzle and increases when it increases. The graphs of the rarefaction ΔP on the axis OZ of the

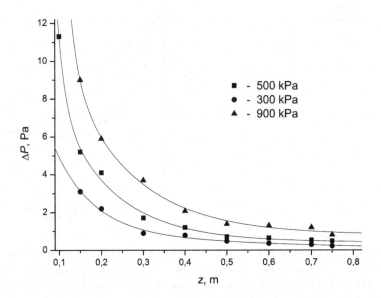

Fig. 2.11 Change of air rarefaction on the axis of the flow at different liquid pressures p in the nozzle

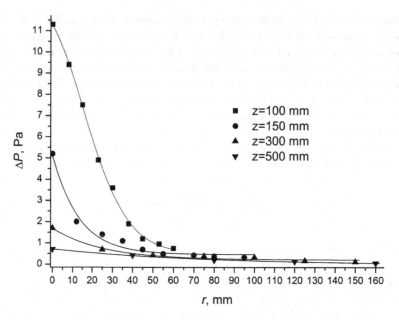

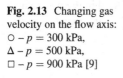

Fig. 2.12 Gas rarefaction profiles at pressure in the nozzle $p = 500$ kPa at different distances z from the nozzle

Fig. 2.13 Changing gas velocity on the flow axis:
○ – $p = 300$ kPa,
△ – $p = 500$ kPa,
□ – $p = 900$ kPa [9]

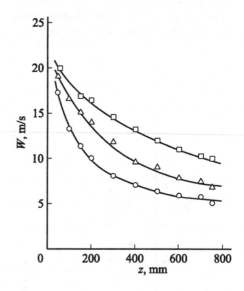

spray flow at the pressure values on the nozzle $p = 300, 500$, and 900 kPa, shown in Fig. 2.11, indicate that the rarefaction decreases with distance from the nozzle to zero value. The air rarefaction ΔP in the two-phase flow decreases from the axis toward the boundary of the flow.

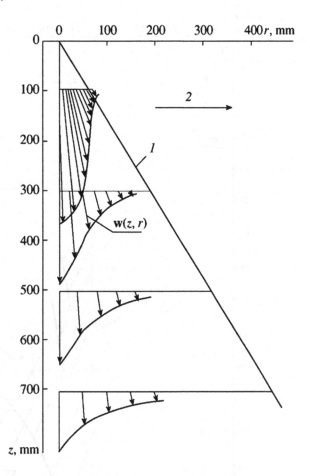

Fig. 2.14 Gas velocity field for the pressure $p = 500$ kPa at the nozzle. (1) The spray boundary and (2) velocity vector scale $|\mathbf{W}| = 10$ m/s [9]

The rarefaction profiles at distances $z = 100$, 150, 300, and 700 mm at the liquid pressure in the nozzle $p = 500$ kPa are shown in Fig. 2.12. It will be recalled that, for measurements at a distance of 100 mm from the nozzle and closer, a shortened static pressure tube with a distance of 30 mm from its front end to the receiving holes was used. In this case, the error of the results increased, and in accordance with the remarks above, they should be considered as estimates.

The graphs of the change in the gas velocity W on the flow axis at different liquid pressures in the nozzle, shown in Fig. 2.13, indicate that the velocity of the gas interacting with the liquid in the two-phase stream is hyperbolically decreasing with distance from the nozzle. This result qualitatively confirms the conclusions made earlier in [15].

Figure 2.14 shows the gas velocity field in the two-phase flow at liquid pressure in the nozzle $p = 500$ kPa. The measurements were carried out in the cross sections of the flow at the distances $z = 100$, 150, 300, 500, and 700 mm from the nozzle outlet.

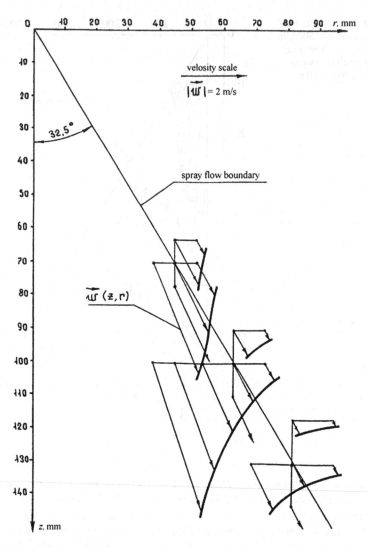

Fig. 2.15 The velocity field of the gas near the conditional boundary of drop stream at a liquid pressure in the nozzle $p = 5 \cdot 10^5$ Pa

The velocity W of the gas in the two-phase flow significantly decrease in the cross section of the flow from the axis to the boundary. In this case, the direction of the velocity vector **W** also changes, as shown in Fig. 2.14. The results in this figure allow us to conclude that the gas motion lines near the flow axis ($r < 200$ mm) differ a little from the rays emanating from the center of the nozzle outlet. It turned out that increasing the pressure of the liquid in the nozzle does not lead to a change in the streamlines but causes only an increase in the modules of the gas velocity vectors.

It is an important fact that the radial components of the gas velocity vectors are directed not to the axis but to the boundary of two-phase flow, which contradicts, as it seems, the fact that the gas flows from the outside into a two-phase flow.

In order to study the conditions for the entry of gas into two-phase flow from the environment, measurements were made for the velocities and static pressures of the gas near the conditional boundary of the stream. The results of these measurements, shown in Fig. 2.15, indicate that the velocity vector of the gas at the boundary of two-phase flow is directed into it at a small angle – about several degrees – to the generatrix of the conditional boundary in the point of its intersection with the velocity vector. The value of this angle decreases with distance from the nozzle. The velocity and static pressure of the gas near the boundary of the drop stream, including the outer region adjacent to it, are monotonously decreased with the increasing distance from the nozzle and from the flow axis. We note that in the outer region adjacent to the flare boundary, the radial components of the gas velocity vectors are directed radially from the axis of the flow (Fig. 2.15).

It should be noted that the above results, relating to the magnitude and nature of the changes in the gas velocity in the sprayed liquid flow along its axis and radius, do not completely correspond to the data of previous studies [13].

This can be explained by the fact that when measuring the gas velocity in a two-phase disperse flow using the Pitot-Prandtl pneumometric tube and the MMN type micromanometer, the authors of [13] did not take measures to eliminate the influence of the sprayed liquid on the functioning of the pressure sensor and, thereby, on the measurement accuracy of micro-pressure by means of a secondary device.

Thus, the Pitot-Prandtl pneumometric tube constructive modernization, designed to eliminate the influence liquid droplets on the measurements of basic hydrodynamic characteristics of gas in a two-phase disperse flow, and the use of a precision micromanometer with a sensitivity of 40 times higher than that of a serial device of the MMN type made possible the experimental study of the pressure and velocity fields of the gas phase in the free two-phase stream produced by a mechanical spray nozzle.

The following conclusions can be drawn, based on the experimental results presented in Sect. 2.2.

1. As a result of the interaction of the sprayed liquid and gas in the free two-phase flow produced by mechanical nozzle, the gas rarefaction of the order of 10 Pa is created, which causes the so-called ejection effect, i.e., the inflow of gas into the drop stream from the outer region.
2. A gas rarefaction in two-phase stream depends on the pressure of the liquid in the nozzle and increases when it is raised.
3. The rarefaction and velocity of the gas in the flow are maximal near the root of spray and decreasing along the axis and radius of the flow to its periphery.
4. The gas flowing lines near the axis of the flow are similar to the rays emanating from the center of the nozzle outlet; their configuration does not depend on the pressure of the liquid on the nozzle.

5. The vector of the velocity of the gas ejected from the outer region can be directed into the two-phase stream under a small – about several degrees – angle to the generatrix of the conditional boundary surface lying with the velocity vector in one plane; the gas velocity modulus at the flow boundary is relatively small – about 1 m/s.
6. The radial components of the gas velocity vectors inside the flare and partly in the outer region adjacent to its boundary are directed from the axis of the flow to its periphery.
7. Considering the insignificantness of the rarefaction of gas in the two-phase flow and the relatively small value of its velocity in comparison with the speed of sound, a change in the gas density and its compressibility in a spray stream can be neglected.

Some of the results presented in this section were previously published in papers [9, 10].

2.3 Experimental Studies of the Spatial Distribution of the Dispersed Phase in the Two-Phase Flow Produced by a Nozzle

Characteristics of the spatial distribution of the disperse phase in the jet of the injector are the distributions of the specific liquid flows $J(r, z)$ and volume concentration of the liquid proportional to its volume fraction $\alpha(r, z)$.

The values of $J(r, z)$ and $\alpha(r, z)$ are functions of the spatial coordinates of the points inside the two-phase flow, and they are related to each other by a directly proportional relationship. For the flow through a horizontally oriented area, the coefficient of proportionality is the axial velocity of the liquid $u_z(r, z)$ at a given point:

$$J(r, z) = u_z(r, z) \cdot \alpha(r, z). \tag{2.11}$$

Due to the difference in the velocities of individual drops in each point of the flow, it is necessary to use in formula (2.11) the droplet velocity averaged on their volume (mass), which, generally speaking, can differ from their velocity averaged on the number of drops, measured by the time-of-flight method.

One of the purposes of the study described below was the experimental evaluation of the effect of the ejection on the formation of the radial distributions of the specific liquid flows and the volume concentration of the dispersed phase in the spray. In particular, it was thus planned to verify the validity of the assumption said in [16] of the straightness of the trajectories of drops in the spray flow, as well as the followed from this assumption and used in [17–19] the conclusion about the self-similarity of the radial profile of the, normalized in the sense of [16].

2.3.1   Measurements of the Volume Concentration of Droplets

To measure the volume concentration of the liquid phase, it was possible to use the same experimental equipment as for measuring the dispersion of the spray by the method of small-angle light scattering. In fact, under the condition of a one-time scattering of light by droplets, the intensity $I(\beta)$ of light scattering by any small angle $\beta < 6°$ is proportional to the number N and, therefore, to the total volume of the droplets which scatter light.

The radial distribution of the volume concentration of the liquid in the spray of the nozzle could be determined by measuring the intensity of light scattering during the propagation of the laser beam along the chords l of spray cross section, remote at different distances l from the axis (Fig. 2.16) with the subsequent use of the Abel transformation [20, 21]. This technique, having limited accuracy, would require a fairly complex numerical processing of the results of measurements of the light scattering signal.

Taking into account that the experimental data on the specific liquid flows and the velocity of the liquid also allow us to determine the volume concentration of droplets with the help of (2.11), its independent measurements by light scattering method can be considered as additional measuring. And in this sense, it seemed reasonable to simplify the measurement procedure.

The simplification was that instead of measuring the signal of light scattering on several chords of the spray flow cross section at several scattering angles in a certain interval [20] (or at least at two angles of light scattering [21]), the measurements

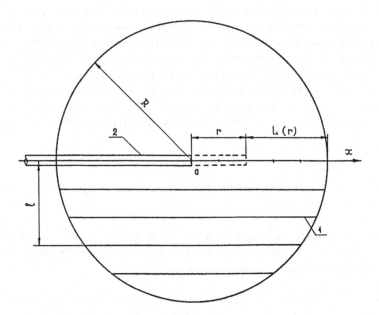

Fig. 2.16 Scheme for measuring the radial profile of volumetric liquid concentration

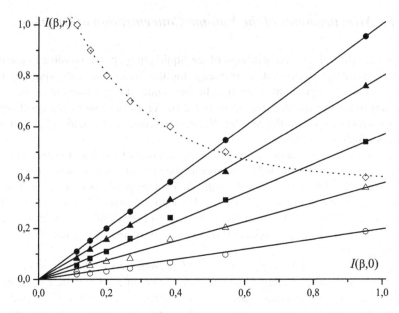

Fig. 2.17 Comparison of the values of the indicatrix of light scattering for different lengths $L(r)$ of a scattering zone along the spray diameter $\bullet - r = 0$ mm; $\blacktriangle - r = 38$ mm; $\square - r = 76$ mm; $\triangle - r = 115$ mm; $\bigcirc - r = 153$ mm; $\Diamond - \beta \cdot 365$ radians

were made only at one angle $\beta_1 \approx 9.5' = 0.00275$ radians when the laser beam go along the diameter of spray. In this case, the length $L(r) = R - r$ of the light scattering zone decreased, beginning with the half diameter of a spray, by five times with uniform step due to the laser beam that passed inside the glass tube 2 in the rest of the diameter of the spray (Fig. 2.16). The conditional boundary of a spray with radius $R = z \cdot \text{tg} \varphi$ was determined from the root angle φ of the spray.

Under the condition of a slight change in the spectral composition of the droplets along the cone radius of the spray, the validity of which was confirmed by a proportional change in the signal $I(\beta, r) \sim I(\beta, 0)$ of light scattering at different angles β with a change in the length $L(r)$ of the scattering zone (Fig. 2.17); we can write

$$I(\beta_1, r) = \text{const} \cdot \int_r^R \alpha(x) \cdot dx = \text{const} \cdot A(r). \qquad (2.12)$$

Thus, by measuring the light scattering signal $I(\beta_1, r)$ only at one angle β_1 at several positions of the glass tube, it was possible to determine the relative integral distribution of the volume fraction of dispersed phase:

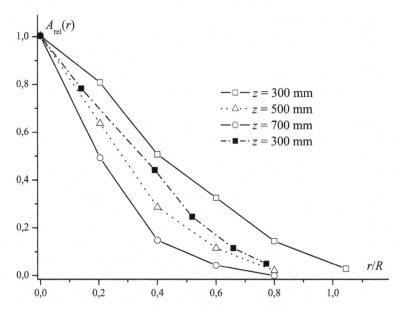

Fig. 2.18 Relative integral concentration profiles fluid in different sections of the spray flow; $P_1 = 5$ atm; ■ – this was obtained using the formulas (2.11) and (2.13)

$$A_{rel}(r) = \frac{A(r)}{A(0)} = \frac{\int\limits_r^R \alpha(x) \cdot dx}{\int\limits_0^R \alpha(x) \cdot dx} = \frac{I(\beta_1, r)}{I(\beta_1, 0)}. \qquad (2.13)$$

The experimental distributions (2.13) obtained in the three cross sections of a spray $z = 300$, 500, and 700 mm at the liquid pressure at the nozzle $P_1 = 5$ atm are shown in Fig. 2.18.

Use of information about the parameters of the measuring device, the focal length f of the lens, the diameter d of the receiving diaphragm, as well as the experimental data about the integral characteristics of the light scattering of droplets, about energy $I_{\Sigma 0}$ of the not scattered laser beam, has allowed to determine approximately the value of the normalization constant $A(0)$ in the distribution (2.13) by formula

$$A(0) = \int\limits_0^R \alpha(x) \cdot dx = \left(\frac{f}{d}\right)^2 \cdot \frac{\lambda}{I_{\Sigma 0}} \cdot \int\limits_0^\infty I(\beta) \cdot d\beta = (6.7 \pm 1.7) \cdot 10^{-6} \ (m) \quad (2.14)$$

A significant error (~25%) in determining $A(0)$ is due mainly to the error in measuring the total energy $I_{\Sigma 0}$ of the laser beam with the help of ten neutral light filters, for each of which the measuring accuracy for double attenuation coefficient was ~2 to 2.5%.

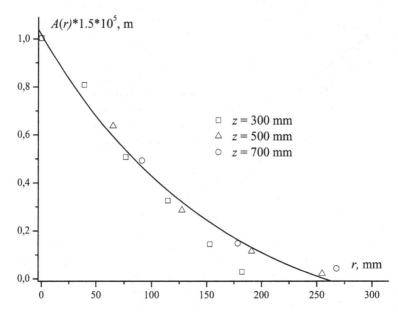

Fig. 2.19 Integral distribution of the volume fraction of droplets by the radius of the spray flow; $P_1 = 5$ atm; the curve is an approximation of experimental data for $z = 500$ mm

By using the already known values of $A_{rel}(r)$ and $A(0)$, the absolute values of the integral distribution $A(r)$ were determined (Fig. 2.19).

By means of numerical differentiation of function $A(r)$, we can approximately obtain the profile of the radial distribution of the liquid volume fraction $\alpha(r)$ in the spray flow; the examples of which are shown in Fig. 2.20 by lighter symbols.

From Figs. 2.18, 2.19, and 2.20, it is obvious that the profile of the volume concentration considered as a function of the dimensionless radius of the spray flow $\rho = r/R$ is visibly deformed (on about 40%) by compressing to the axis on increasing the distance from the nozzle. Conversely, if this profile is considered as a function of the absolute radius r, its relative widening at the $z = 300$–700 mm section equals the same value, about 40%.

The observed character of the change in the volume concentration profile of the liquid in the measured range of heights z of two-phase flow is intermediate between their movement along rectilinear trajectories of type of the rays emanating from the center of the outlet hole of the nozzle, as was assumed in [16], and the almost vertical drop of the main mass of droplets at a constant rate. The cause causing the curvature of the trajectories of droplets toward the axis of the torch is the interphase interaction, which causes the gas flow from the outer region into the spray cone.

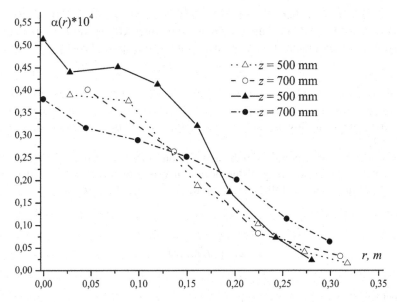

Fig. 2.20 Profiles of the volume fraction distribution of liquid; Δ, ○ – obtained by differentiating the integral profile $\alpha(r) = dA(r)/dr$; ▲, ● – from the data about the specific liquid flows and the velocity of droplets $\alpha(r) = J(r)/u_z(r)$; $P_1 = 5$ atm

2.3.2 Measurements of Distribution of the Specific Liquid Flows

The method for measuring the distribution of specific liquid flows over the cross section of the spray flow is based on the use of cylindrical tube probes of relatively small diameter (in comparison with the transverse dimensions of the spray flow). In the measurements, two 150 mm long tubes were used, with a wall thickness of 1 mm, with a sharp leading edge and a fitting at the lower end, to which a flexible rubber hose was connected to discharge the liquid taken from the stream into a measuring container, the beaker. The internal diameters of the sampling tubes were 7.70 ± 0.05 mm and 18.8 ± 0.1 mm. Near nozzle at $z = 100$ mm, a tube of smaller diameter was used for the measurements; at larger distances $z = 300$ mm, both tubes are used. At a considerable distance $z = 500$–700 mm from the nozzle, the measurements were carried out with a tube of a larger diameter.

In each individual experiment, a time was measured using a stopwatch, during which a predetermined volume $V = 20$–100 ml of a liquid withdrawn by a vertically oriented sampler placed at a given "point" of the stream flowed into the measuring container. The coordinates of the measurement point were determined by the position of the center of the upper section of the probe tube. The value of the specific flow of liquid at a given point was calculated by dividing a given volume of liquid by the time it flowed and by the value of cross-sectional area of the tube with an accuracy of not worse than 5%. An experimental estimate of the effect of gas flowing conditions in the vicinity of the sampling tube, which consisted of comparing the experimental data obtained at the same points of the spray flow by tubes of two different diameters differing by more than twofold, showed that the difference in the experimental results in this case does not exceed 10%. Moreover, for a tube with a larger diameter, the measured values of the specific liquid flows turned out to be everywhere lower, as expected. Therefore, special measures to ensure the isokinetic conditions of the gas flowing during the sampling of the fluid were not undertaken in this study.

In work [16], proceeding from the assumption of the straightness of the trajectories of the drops in the spray of the nozzle, it was shown that when used for the normalized distribution of the specific liquid flows $J(r, z)$ of formula

$$J_n(\rho, z) = \frac{J(r,z)\pi R^2}{2\pi \int\limits_0^{R(z)} J(r,z)rdr}; \quad \rho = r/R \tag{2.15}$$

so that

$$\int\limits_0^1 J_n(\rho, z)\rho d\rho = \frac{1}{2},$$

the normalized profiles $J_n(\rho, z)$ as functions of the dimensionless radius ρ turn out to be self-similar in the cone height z of the spray flow. Note that under this assumption, the side boundary of the spray flow at any height z is the surface of the cone defined by the root angle φ, with a radius $R = R(z) = z \cdot \text{tg}(\varphi/2)$. Then from (2.15), taking into account the conservation of the fluid flow, it follows that for the self-similarity of $J_n(\rho, z)$, the function of type

$$J(r,z) \cdot z^2 \sim J_n(\rho, z) \tag{2.16}$$

must be also self-similar in the altitude z of the spray flow.

Figure 2.21 shows the profiles of the specific liquid flows in the form (2.16), obtained with use of sample tubes in several cross sections of the spray flow at different distances $z = 100, 300, 500,$ and 700 mm from the nozzle at a water pressure $P_1 = 5$ atm.

The obvious difference in the configuration of the curves corresponding to different z in Fig. 2.21 indicates a change in the normalized profile of the specific liquid flows $J_n(\rho, z)$ over the spray flow height and, therefore, to the incorrectness of the assumption of the straightness of the droplet trajectories in the general case.

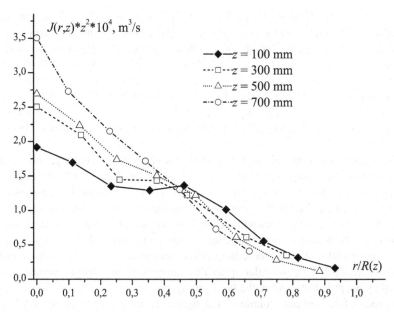

Fig. 2.21 Radial distributions of the specific liquid flows in the spray flow of nozzle; $P_1 = 5$ atm

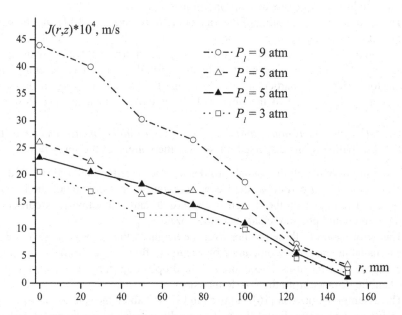

Fig. 2.22 Radial profiles of specific liquid flows at various pressures in the nozzle; $z = 300$ mm

Figure 2.22 shows the profiles of the specific liquid flows measured in the single cross section of the spray flow ($z = 300$ mm) at different water pressures in the nozzle $P_1 = 3$, 5, and 9 atm.

Obviously, the shape of the profiles with a change of pressure in above limits remains the same to a large extent. Figure 2.22 also indicates the anisotropy of the distribution of the specific liquid flows over the azimuthal angle of the spray flow: data for dashed line and solid line were obtained by measuring $J(r, z)$ in two of its mutually perpendicular axial sections. The above circumstances indicate a significant effect of design parameters and features of the nozzle determining the initial distribution of the liquid flow upon exit from the nozzle on the distributions of velocity and concentrations and of specific liquid flows over the all volume of the spray flow.

The experimental data about the specific liquid flows, its velocity, and volume concentration are compared with each other using formula (2.11). Examples of comparison are shown in Figs. 2.18 and 2.20, where the respective profiles were calculated from the data of $J(r, z)$ and $u_z (r, z)$ using (2.11) and (2.13). The observed difference between the profiles obtained by different methods is explained, firstly, by the aforementioned anisotropy of the spray flow on the azimuthal angle: for convenience, the profiles of the specific liquid flows and its concentration were measured over two mutually perpendicular spray flow diameters. Secondly, a certain role is played by the polydispersity of the liquid spray, due to which there are differences in the speeds of movement, volumes, and light scattering properties of droplets of different diameters, and therefore their contributions to the specific liquid flow and signal of light scattering can be disproportionate.

The accuracy in determining of the value of $A(0)$ by formula (2.14) and the scale of dashed and dots curve along the ordinate axis in Fig. 2.20 is limited. Taking this into account, the correspondence of profiles $J(r, z)$ and $\alpha(r, z)$ obtained by independent methods, which was established by the formula (2.11), can be considered satisfactory. Hence, in particular, it follows that the average velocity of drops u_z, measured by the method of the time-of-flight, is close to their average flow rate velocity.

The following conclusions can be drawn on the results of measurements of the spatial distribution of the dispersed phase over the volume of the spray flow:

1. The initial shape of the radial profiles of the specific liquid flows and its concentrations, in particular, their configuration in the root zone of the flow, is largely determined by the design of the nozzle and, accordingly, is conserved when the liquid pressure changes in the nozzle.
2. The deformation of these profiles over the height of the spray flow is caused by interfacial interaction accompanied by suction (inflow) of gas from the surrounding space, which causes deformation of the droplet trajectories to the axis of flow when the distance from the nozzle increases.
3. The assumption made in [16] and used in [17–19] about the self-similarity of the profile of the specific liquid flows is generally not valid for a larger part of the spray flow, except for its near root zone, with all consequences resulting from this assumption.
4. In particular, the measurement of the relative distribution of the specific liquid flows (2.15) in only one cross section of the spray flow and its propagation to any

other cross section [22] should be considered incorrect, and attempts to create nozzles with the same distribution of the specific liquid flows in any cross section of the spray flow are not justified.

References

1. Rusanov, A. A., et al. (1969). *Ochistka dymovykh gazov v promyshlennoy energetike (Cleaning of flue gases in industrial power engineering)*. Moscow: Energia.
2. Shifrin, K. S., & Golikov, V. I. (1961). Determination of the droplet spectrum by the method of small angles. In *Investigation of clouds, precipitation and thunderstorm electricity. Proceedings of the sixth interdepartmental conf.* (Izd. AN SSSR, Moscow), pp. 266–277.
3. Shifrin, K. S., & Kolmakov, I. B. (1967). Calculation of the particle size spectrum from the current and the integrand values of the indicatrix in the region of small angles. *Izvestiya AN SSSR. Fizika Atmosfery i Okeana (Physics of the Atmosphere and the Ocean), 3*(12), 1271–1279.
4. Bayvel, L. P., & Lagunov, A. S. (1977). *Measurement and control of the dispersion of particles by the light scattering method*. Moscow: Energia.
5. Dieck, R. H., & Roberts, R. L. (1970). The determination of the sauter mean droplet diameter in fuel nozzle sprays. *Applied Optics, 9*, 2007–2014.
6. Zimin, E. P., & Krugersky, A. M. (1977). Integral characteristics of light scattering by polydisperse particles. *Optika i Spektroskopiya (Optics and Spectroscopy), 43*(6), 1144–1149.
7. Zakharchenko, V. M. (1975). Measurement of the flow velocity by a laser one-beam time-of-flight method. *Uchenyye zapiski TsAGI (Scientific Notes of CAHI), 6*(2), 147–157.
8. Zhigulev, S. V. (1982). On one version of the laser single-beam time-of-flight method for measuring the flow velocity. *Uchenyye zapiski TsAGI (Scientific Notes of CAHI), 13*(5), 142–147.
9. Simakov, N. N. (2004). Crisis of Hydrodynamic Drag of Drops in the Two-Phase Turbulent Flow of a Spray Produced by a Mechanical Nozzle at Transition Reynolds Numbers. *Zhurnal Tekhnicheskoj Fiziki, 74*(2), 46. [Tech. Phys. 49, 188 (2004)].
10. Simakov, N. N., & Simakov, A. N. J. (2005). Anomaly of gas drag force on liquid droplets in a turbulent two-phase flow produced by a mechanical jet sprayer at intermediate Reynolds numbers. *Applied Physics, 97*, 114901.
11. Kremlevsky, P. P. (1980). *Measurement of flow and quantity of liquid, gas and steam*. Moscow: Izd-vo Standartov.
12. Povkh, I. L. (1969). *Technical hydromechanics*. Leningrad: Mashinostroyeniye.
13. Katalov, V. I., et al. (1975). Experimental determination of the velocity of a continuous phase in a dispersed flow of a free spray flow of a mechanical injector. In *Massoobmennye i teploobmennyye protsessy khim. tekhnol. (Mass exchange and heat exchange processes of chemical technology)* (YaPI, Yaroslavl), pp. 13–16.
14. Leidenforst, W., & Ku, J. (1960). *New high-sensitivity micromanometer*. Transl. from English, Instruments for scientific research, No. 10, pp. 76–78.
15. Gelperin, N. I., et al. (1974). Spraying liquid with mechanical injectors. *Teor Osnovy Khim Tekhnol (Theory Fundamentals of Chemical Engineering), 8*(3), 463–467.
16. Aniskin, S. V. (1978). Similarity of the density of irrigation fluid sprayed by a mechanical injector SGP. In *Protection of the environment from pollution by industrial emissions in pulp and paper industry* (LTA, LTITSBP, Leningrad), No. 6, pp. 165–168.
17. Mikhailov, E. A., et al. (1981). *Development of a methodology for calculating the geometric dimensions of nozzles with a given character of the distribution of specific fluid flows* (Ruk. dep. ONITEKHIM, 20.04.1981, Yaroslavl), p. 6.

18. Mikhailov, E. A. (1982). Dissertation, Moscow Institute of Fine Chemical Technologies named after M.V. Lomonosov.
19. Pazhi, D. G., & Galustov, V. S. (1984). *Fundamentals of spraying technology*. Moscow: Khimiya.
20. Zimin, E. P., et al. (1973). Optical measurements of the parameters of the dispersed condensed phase of two-phase flows. *Teplofizika Vysokikh Temperatur (Thermal Physics of High Temperatures), 11*(15), 1037–1043.
21. Zimin, E. P., et al. (1975). Determination of the volume concentration of a dispersed phase from the measurement of the scattering of light at two angles. *Optika i Spektroskopiya (Optics and Spectroscopy), 39*(1), 155–161.
22. Pazhi, D. G., & Galustov, V. S. (1979). *Sprayers of liquid*. Moscow: Khimiya.

Chapter 3
Analyses of Experimental Results: Physical Picture of a Free Two-Phase Flow Generated by a Mechanical Nozzle

An analysis of the experimental data for a two-phase spraying stream showed that it has a number of features. In particular, such a flow is strongly turbulent; the average velocities of phases differ markedly at all points of the flow; there is a noticeable rarefaction of gas in the root zone of the spray; the hydrodynamic drag coefficient of each drop is smaller than of a single spherical particle at the same value of the Reynolds number.

3.1 Analogy of Turbulent Motion of Gas in a Spray Flow with a Flooded Jet

An estimate of the Reynolds number Re, which characterizes the degree of turbulence of the gas, in average cross section of two-phase stream with a diameter of $D \approx 400$ mm (at height $z = 300$ mm) gives

$$\mathrm{Re} = \frac{w \cdot D \cdot \rho_g}{\mu_g} = \frac{10 \cdot 0.4 \cdot 1.2}{1.8 \cdot 10^{-5}} = 2.7 \cdot 10^5 \gg \mathrm{Re_{cr}} = 2.3 \cdot 10^3 \qquad (3.1)$$

Relation (3.1) indicates the developed turbulence of the gaseous flow in the spraying stream.

However, in one of the phenomenological approaches [1], which are applied to the modeling of the hydrodynamics of a nozzle spray and described in the first chapter, the turbulence of a gas, in particular, the turbulent transfer of a pulse inside a continuous phase is not taken into account. In this connection, it seemed interesting, using experimental data, to find out: how, namely, and to what extent does the turbulence of the two-phase flow in the spray of a mechanical injector manifest itself?

© Springer Nature Switzerland AG 2020
N. N. Simakov, *Liquid Spray from Nozzles*, Innovation and Discovery in Russian Science and Engineering, https://doi.org/10.1007/978-3-030-12446-5_3

To this end, an attempt was made to find a possible analogy between the flow of gas in the spray of nozzle and the turbulent flooded jet [2]. Note that for the pneumatic spraying of liquids, such an analogy was established in the works of G.N. Abramovich et al. [3–7], while in the literature, there is no information on the existence of this analogy for the spray flows created by mechanical injectors.

To reveal the abovementioned analogy, in interpreting experimental data on the distribution of gas velocities in the two-phase flow produced by a nozzle, certain methodological elements of the theory of turbulent jets were used.

Figure 3.1 shows graphs of the dependence of the quantity $1/w_m$, inverse to the gas velocity on the flow axis, on the axial coordinate at three pressures on the nozzle: $P_1 = 3, 5, 9$ atm.

It is evident from the graphs that this dependence at all pressures is linear:

$$\frac{1}{w_m} = \frac{z + Z_0}{\alpha} \tag{3.2}$$

The values of Z_0 and a in (3.2) depend on the liquid pressure P_1 in the nozzle, as it is reflected in Table 3.1.

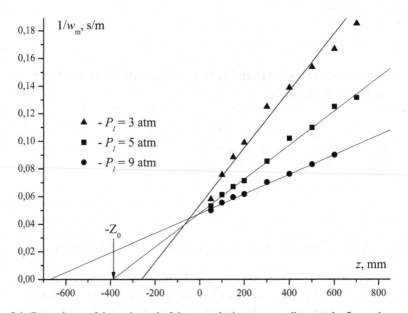

Fig. 3.1 Dependence of the reciprocal of the gas velocity on a coordinate at the flow axis

Table 3.1 The values of Z_0 and a as a function of the liquid pressure P_1 in the nozzle

P_1, atm	Z_0, mm	a, m²/s	b = a/P₁, m²/(s·atm)
3	260	4.7	1.57
5	390	8.0	1.60
9	670	13.9	1.54

It is also evident from the table that the ratio

$$a(P_1)/(P_1) = b \cong (1.57 \pm 0.03) \ \text{m}^2/(\text{s} \cdot \text{atm}) \tag{3.3}$$

does not depend on the pressure with an accuracy of 2%. This means that the function $a = a(P_1)$ is a directly proportional relationship:

$$a(P_1) = b \cdot P_1. \tag{3.4}$$

From (3.2), taking (3.4) into account, it follows that

$$w_m = \frac{\alpha}{z + Z_0} = \frac{b \cdot P_1}{Z}. \tag{3.5}$$

Thus, according to the obtained experimental data (Fig. 2.13), the velocity of the gas on the axis of the spray flow of the mechanical nozzle decreases in inverse proportion to the coordinate $Z = z + Z_0$, measured from a certain pole located above the nozzle at a distance Z_0 (Fig. 3.1). We note that the velocity of the gas on the axis of the flooded turbulent jet varies in a similar manner [2].

Figure 3.2 shows the profiles of the dimensionless gas velocity $W = w/w_m$ as a function of the dimensionless coordinate $\theta = r/Z$, obtained from the velocity measurements (Fig. 2.14) in different cross sections of the spray flow $z = 100$,

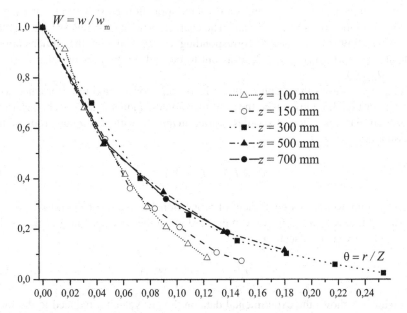

Fig. 3.2 The radial profiles of the dimensionless gas velocity in two-phase flow; $P_1 = 5$ atm

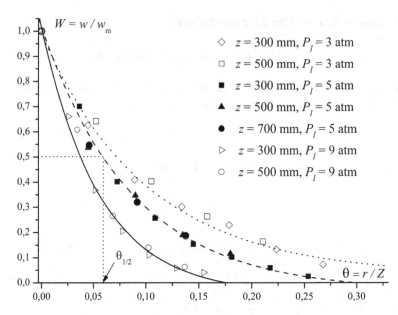

Fig. 3.3 The radial profiles of the dimensionless gas velocity in the spray stream at various liquid pressures in the nozzle

150, 300, 500, 700 mm at liquid pressure in the nozzle $P_1 = 5$ atm. It is obvious that for $z > 300$ mm, the radial profiles of the gas velocity are practically self-similar.

Figure 3.3 indicates that this self-similarity takes place also at other liquid pressures in the nozzle ($P_1 = 3$ and 9 atm) at the spray flow cross sections sufficiently remote from the nozzle: $z > 300$ mm. The characteristic dimensionless radius of the two-phase flow $\theta_{1/2} = r_{1/2}/Z$, corresponding to the value of the dimensionless velocity $W = w(r_{1/2})/w_m = 1/2$, turns out to depend on the pressure at the nozzle $\theta_{1/2} = \theta_{1/2}(P_1)$.

Figure 3.4 shows the graph of this relationship, constructed in accordance with the data in Fig. 3.3. Obviously, the jet dimensionless radius, which is characteristic of self-similar gas velocity profiles, varies inversely with the square root of the pressure at the nozzle

$$\theta_{1/2}(P_1) = C_1/(P_1)^{1/2}, \tag{3.6}$$

and according to Fig. 3.4 coefficient of proportionality $C_1 = 0.133 \ (\text{atm})^{1/2}$.

Taking into account (3.6), we can proceed to another dimensionless radius of the torch in its given cross section:

$$\psi = r/r_{1/2} = r/(Z \cdot \theta_{1/2}). \tag{3.7}$$

Figure 3.5 shows the experimental data on the gas velocity, reduced to the form $W = f(\psi)$.

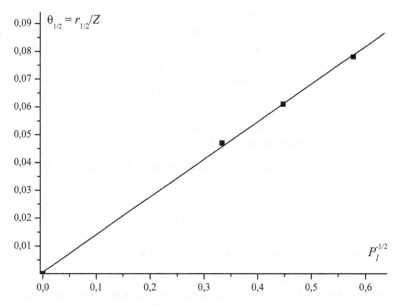

Fig. 3.4 Dependence of the characteristic dimensionless radius of the gas jet in the spray flow on the liquid pressure P_l in the nozzle

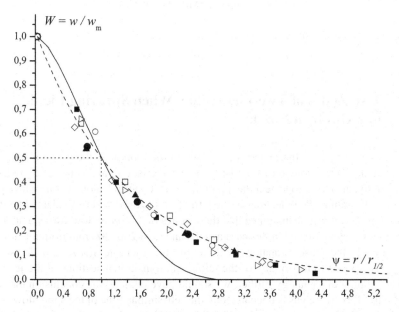

Fig. 3.5 Universal profiles of the dimensionless gas velocity in the liquid jet sprayed by a nozzle; the notations are the same as in Fig. 3.3; dashed curve is graph of function $W = \exp(-\psi \cdot \ln 2)$; solid cure is the Tolmin profile for an axisymmetric turbulent flooded jet

Obviously, all the experimental points are located near the same curve. This means that in the spray flow cross sections sufficiently remote from the nozzle ($z > 300$ mm), the dimensionless velocity profiles of the gas obtained at different pressures on the nozzle are self-similar. This property of the spatial distribution of the velocities of the continuous phase is also characteristic of a flooded turbulent jet [2]. Note that curve in Fig. 3.5 is very similar to the graph of exponential dependence.

Thus, the analogy of gas flow in the free spray flow produced by a mechanical nozzle and a flooded turbulent jet is as follows:

1. The velocity of the gas at the axis of the flow varies inversely with the axial coordinate

$$w_m \sim 1/Z \qquad\qquad (3.8)$$

also, as in the main region of an axisymmetric flooded jet.

2. The radial profiles of the dimensionless gas velocity as a function of the dimensionless radius of the flow, starting from some distance from the nozzle, are practically self-similar in altitude of the spray flow

$$W = f(\psi) = \exp(-\psi \cdot \ln 2) \qquad\qquad (3.9)$$

at all created pressures in the nozzle.

3.2 Two Zones of Two-Phase Flow When Spraying Liquid in a Gas by a Nozzle

The above analysis of the experimental data on the motion of the gas in a spray of nozzle makes it possible to distinguish two characteristic spatial zones of two-phase flow: firstly, an active zone near the top of the spray cone and another zone where gas flow is self-similar. The meaning of the term "active zone" will be explained below. Schematically, it can be imagined that these zones of the spray flow are separated by a transition region located in this case at a distance $z \approx 200$–300 mm from the nozzle (Fig. 3.2). In the sense of analogy with the turbulent jet, the first zone of the spray flow corresponds to the totality of the initial and transitional sections of the jet.

Developing this analogy further, we should have come to the conclusion that the total pulse flux of the gas is unchanged in the self-similar zone of the spray flow over its height [2]. Hence, in turn, it would follow that the total pulse flux of the liquid in the same zone of the spray flow is also practically preserved. Taking into account the existing difference in the velocities of two phases and their dynamic interaction in the flow, the statements about the conservation of the momentum flux of each phase in a large area of the spray flow seem paradoxical and deserve special consideration (see Sects. 3.3 and 3.4).

As regards the motion of the gas phase, then, as follows from (3.5), (3.7), and (3.9), in the flow cross section given by the axial coordinate $Z = z + Z_0$, the dependence of the gas velocity on the radius r has the form

$$w_z(r) = w_m \cdot W = \frac{b \cdot P_1}{Z} \cdot f\left(\frac{r \cdot \sqrt{P_1}}{C_1 \cdot Z}\right) = \frac{b \cdot P_1}{Z} \cdot f(\psi) \qquad (3.10)$$

The momentum flux of an incompressible gas through a given cross section of the spray flow is determined by the integral

$$L_g = \int_0^\infty \rho_g \cdot w_z^2(r) \cdot 2\pi \cdot r \cdot dr = 2\pi \cdot \rho_g \cdot b^2 \cdot C_1^2 \cdot P_1 \cdot \int_0^\infty f^2(\psi) \cdot \psi \cdot d\psi \qquad (3.11)$$

It is obvious from (3.11) that in the self-similar zone, the pulse flux of the gas is practically independent of the axial coordinate, and its magnitude is directly proportional to the pressure in the nozzle. Note that this conclusion was obtained on the basis of an analysis of the experimental data on the motion of gas in a spray flow. The correspondence between the statement about the conservation of the impulse flux of the disperse phase in the self-similar zone of the spray flow and the experimental data will be considered below. Here we will explain the meaning of the term "active zone" near the root of the spray flow. The essence is the following.

The conservation of the gas-pulse flux in the self-similar zone of the spray flow means that, in spite of the visible difference in the velocities of phases in this zone (see Chap. 2), their force interaction, which causes momentum interphase exchange, is insignificant. At the same time, the movement of the gas as a whole is due to the motion of the dispersed liquid and its force action on the gas. Consequently, the transfer of mechanical energy and momentum from the liquid to the gas occurs and almost ends in the root zone. In this sense, it is called active: in the root zone, there is an intensive dynamic interaction of the phases, and taking into account the analogy of the transfer processes, apparently, there are the most intense interphase heat and mass transfer.

The reason for this phenomenon can be partly a significant difference in the mean volume concentrations of the liquid in the root and self-similar zones of the spray flow (see Chap. 2).

3.3 Features of Motion of a Dispersed Phase in a Free Spray Flow

The purpose of the below following analysis of experimental data on the motion of a dispersed liquid in the flow from nozzle, as well as when considering the motion of a gas, mainly consists in presenting a qualitative physical picture of the two-phase flow under study. Involvement of some mathematical relations and quantitative data is intended to serve this goal, facilitating, to a certain extent, theoretical ideas.

One of the features of a two-phase flow in a free spray jet of a mechanical injector is that the effects of secondary crushing and coagulation of droplets in the spraying stream are not as significant as, for example, in high-speed two-phase flows of Laval nozzles [8]. Indeed, according to the experimental data given in the second chapter, the change in the average droplet diameter d_{32} over a sufficiently long segment (0.6 m) of the spray flow is only about 10%. This experimental fact is due to the low probability of collision of droplets.

Returning to the experimental data on droplet velocities (Fig. 2.8), we note that all the profiles of the axial velocity component are externally similar. After reduction to the dimensionless form, it turned out that these profiles at a sufficient distance from the nozzle are self-similar (Fig. 3.6) by analogy with the gas velocity profiles (Figs. 3.2 and 3.3). In Fig. 3.6 designation u_{zm} indicates the droplet velocity on the axis of the spray stream in a given cross section.

Self-similarity of the liquid velocity profile means that for its description in the known zone of the spray flow analogous to the velocity profile of the gas (3.9), one universal function can be used. The specific form of this function can be determined by an analytical approximation of the experimental data. In the simplest case, with a precision of ±15% of the average velocity over the cross section of flow, a function of a constant value can be used as an approximating function.

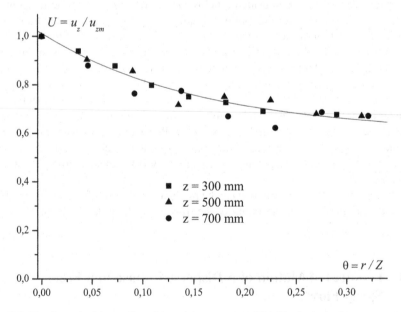

Fig. 3.6 The dimensionless profiles of the axial component of liquid velocity in the spray stream at liquid pressure in the nozzle $P_1 = 5$ atm

The profiles of the volume fraction distribution of liquid obtained by the light scattering method in different cross sections of the two-phase flow (Figs. 2.19 and 2.20) in the self-similar zone at $z \geq 300$ mm differ insignificantly and can be roughly approximated by using the linear function

$$\alpha(r) = \alpha_0 - k \cdot r \qquad (3.12)$$

The parameters α_0 and k can be determined from the experimental data (see Sect. 2.3).

Let us now consider the experimental data on the distribution of the specific fluxes of the dispersed phase. Note that, according to Figs. 2.21 and 2.22, the profiles of specific flow liquid $J(r, z)$ for the investigated VTI-type nozzle can also be approximated by a linear function of the radius r of the measurement point

$$J(r,z) = J_0(z) - j(z) \cdot r. \qquad (3.13)$$

The parameters J_0 and j in (3.13) depend on the coordinate z of the given cross section of the spray flow and are related to each other by the characteristic radius $r_0 = r_0(z)$ of the flow at which both sides of Eq. (3.13) vanish. It follows from (3.13) that the radial distribution of the liquid mass flow $J \cdot r$ (i.e., the quantity proportional to the flux through an infinitesimally narrow ring of radius r) should in that case be described by a parabolic dependence

$$J \cdot r = J_0 \cdot r - j \cdot r^2. \qquad (3.14)$$

The experimental curves in Fig. 3.7, constructed from the same data as the curves in Fig. 2.21, really look like inverted parabolas.

The areas under the curves in Fig. 3.7 are obviously approximately equal. By calculating these areas by integration, for the total volumetric flow rate of the liquid in the spray flow at a pressure in the nozzle $P_1 = 5$ atm, the value $V_1 \approx 7.0 \cdot 10^{-5}$ m^3/s was obtained. Direct measurements of the flow of liquid through the nozzle, the flow rate coefficient of which turned out to be equal to $k_V = 0.75$, gave the value $V_1 \approx 7.5 \cdot 10^{-5}$ m^3/s. A satisfactory (with an accuracy of 7%) coincidence of these two values of the liquid flow rate in the spray stream indicates the correctness of the experimental results obtained for the liquid specific flows, in spite of the fact that the isokineticity conditions were not specially created for measurements with sample tubes.

If the total volume flow rate of the dispersed phase is preserved along the axis of spray flow, then from relation (3.13), it follows formula

$$J_0(z) = j(z) \cdot r_0(z). \qquad (3.15)$$

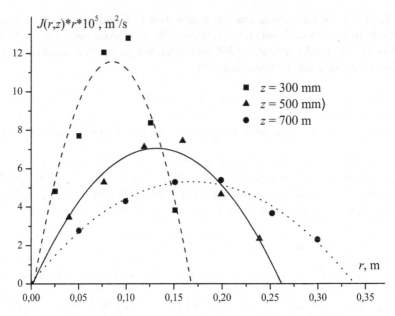

Fig. 3.7 Distribution of the volumetric liquid flow on the radius of the spray stream in its different cross sections; $P_1 = 5$ atm

It connects the parameters of linear approximation for the profiles of the specific liquid flows. Integrating (3.14) over r in the range from 0 to r_0, one can get the relation for mass flow rate of liquid

$$G_1 = \pi/3 \cdot J_0(z) \cdot r_0^2(z) \cdot \rho_1 \qquad (3.16)$$

If at least one of the three functions appearing in Eqs. (3.15) and (3.16) is known, it is possible to determine the other two.

Thus, for a mathematical description of the distribution of the specific liquid flows in the spray flow of the VTI-type nozzle using the approximation linear in radius r (3.13), it needs to find one of the functions $J_0(z)$, $j(z)$, or $r_0(z)$ of the axial coordinate z.

With the specification of the type of these functions, the situation is somewhat more complicated. In particular, taking into account that at the initial part of the spray flow, the trajectories of the drops are close to rectilinear rays emitted from the outlet hole of the nozzle, it would be highly desirable to use a directly proportional dependence:

$$r_0 = \text{const} \cdot z \qquad (3.17)$$

However, the curvature of the droplet trajectories during the transition from the active zone near the root of a spray flow to its self-similar zone does not allow us to

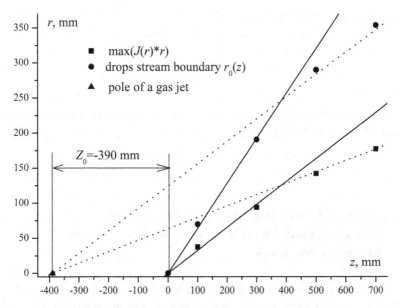

Fig. 3.8 The change in the conditional boundary of the drip stream when moving away from the nozzle according to the data on the specific flows of the liquid

restrict by this simple approximation. But at the same time, one can use the approximation in the form of a piecewise linear function that in both zones has the same form (3.17). The difference between them is in that, in the self-similar zone, the role of the independent variable is played not by the distance to the nozzle z but by the distance to the pole of the turbulent gas jet $Z = z + Z_0$ (see Sect. 3.1). It means that in the self-similar zone, the spray flow expands according to the rays emanating from the pole, and not from the outlet of the nozzle, as in the first zone (Fig. 3.8).

As was shown above, the change in the flow of a gas momentum in the self-similar zone of the free spray flow (practically) is negligible. Hence it follows that the flow of a liquid momentum in the indicated zone also changes in an insignificant measure. Returning to the experimental results shown in Figs. 3.5, 3.6, and 3.7 and taking into account that the dimensionless radial profiles of the axial velocity of the droplets are also self-similar and the change in their velocity U_{zm} along the flow axis is insignificant (see Chap. 2), we come to the conclusion that the approximate conservation of the momentum flux of the disperse phase in the self-similar zone of the free spray flow is in agreement with experimental data.

From the facts of simultaneous conservation by the height of the spray stream in this zone of mass flow of liquid

$$G_1 = 2\pi\rho_1 \int_0^{r_0} J(r,z)\,rdr \qquad (3.18)$$

as well as its flow of impulse

$$L_1 = 2\pi\rho_1 \int\limits_0^{r_0} U_z J(r,z)\, r dr \tag{3.19}$$

it follows that the liquid velocity, averaged over cross section of the spray stream, changes insignificantly [2, p. 30]:

$$< U_z >= L_1/G_1 \approx \text{const.} \tag{3.20}$$

3.4   Crisis of Hydrodynamic Drag for Droplets in a Turbulent Two-Phase Flow at Transitional Reynolds Numbers

When processing experimental data about the hydrodynamics of a two-phase flow in a spray of a mechanical nozzle, an anomaly of the magnitude of the droplet hydrodynamic drag was found: the drag coefficient was found to be four to seven times lower than the known values. For its explanation, a number of hypotheses were involved. The analysis of them led to the conclusion that in a strongly turbulent flow, there can occur an "early" drag crisis when a gas flows around droplets already at transitional Reynolds numbers Re > 50. Thus, a new physical phenomenon was discovered, which explains the features of the two-phase flow that were found in the experiment, in particular, the limited momentum transfer from a liquid to a gas: approximately half of what was originally present in the liquid jet.

It is known the phenomenon of sharp and significant (three to four times) decrease in the drag coefficient C_d of the ball, cylinder, or other poorly streamlined body, arising at Reynolds numbers Re of the order of 10^5. It is called the drag crisis [9–11] and explained by the laminar boundary layer separation from the surface of the streamlined body, transformation into a turbulent boundary layer with displacement of the separation line downstream into the aft region. This changes the pressure profile and improves the flow around a body, approaching the ideal one [9, 10].

In the above-described experimental study of the hydrodynamics of a two-phase spray flow produced by a centrifugal jet nozzle VTI-type (designed by the All-Russia Institute of Heat Engineering) with an outlet hole diameter of 2 mm, the drag crisis was observed for transitional Reynolds numbers, Re $\approx$ 40–130 [12].

Figures 2.7, 2.8, 2.13, and 2.14 show, respectively, the radial profiles of the axial component $U_z(r,z)$ of the mean velocity of the drops, the gas velocity field $\mathbf{W}(r,z)$ at $p = 500$ kPa, and the variation of these quantities with the coordinate z at the axis of flow (at $r = 0$) for liquid pressure in a nozzle $P_1 = 300$, 500, and 900 kPa.

Figure 3.9 shows the change in the relative velocity of phases on the axis of the spray flow at liquid pressure in the nozzle $P_1 = 5$ atm. Obviously, in the self-similar

Fig. 3.9 Approximation of phase velocities and droplet acceleration at the spray flow axis: 1, mean speed of drops U_z; 2, gas velocity W; 3, relative phase velocity $W_{rel} = U_z - W$; 4, drop acceleration with the opposite sign $-0.1 \cdot a_z$ at liquid pressure in the nozzle $P_1 = 500$ kPa [13]

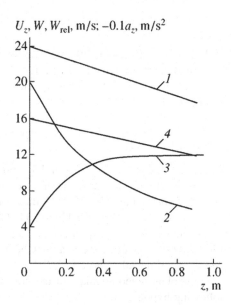

U_z, W, W_{rel}, m/s; $-0.1a_z$, m/s^2

zone of the flow, the relative velocity of phases asymptotically tends to a steady value of approximately 12 m/s.

This feature of the relative phase motion in the turbulent two-phase flow of the spray is not consistent with the description of the interfacial interaction with the aid of formulas (1.4) and (1.7). Thus, in calculating the droplets of the average diameter $d_{32} = 140$ μm moving near the conditional boundary of the flame, where the gas velocity is approximately 1 m/s and less, it turned out that these droplets should be braked to a gas velocity at distances 300–500 mm from the nozzle.

The available experimental data made it possible to estimate the value of the drag coefficient C_d of drops in the spray flow. The following considerations were used. Firstly, in the neglect of the action of gravity, the acceleration of droplets is entirely due to the force of the interphase interaction (1.3). Secondly, the acceleration can be calculated from the experimental data on the droplet velocity using formula

$$a_z = \frac{du_z}{dt} = \frac{du_z}{dz} \cdot \frac{dz}{dt} = u_z \cdot \frac{du_z}{dz} = \frac{1}{2} \cdot \frac{du_z^2}{dz} \qquad (3.21)$$

Thirdly, in calculating the acceleration of droplets according to this formula, one can use the linear approximation of the experimental data (Fig. 3.9) for drops velocity on flow axis:

$$U_z = U_{z0} - v \cdot z \qquad (3.22)$$

At pressure in the nozzle $P_1 = 5$ atm, these parameters of the function (3.22) have the values $U_{z0} = 24$ m/s, $v = 6.7$ s^{-1}. To calculate the relative phase velocity on the flow axis, in addition to (3.22), one can also use the approximation (3.5) for gas velocity.

Using the data of Fig. 3.9, knowing the mean diameter $d = d_{32} = 140$ μm of the drops and the hydrodynamic drag force $F = ma_z$, and neglecting the force of gravity mg, one can determine the drag coefficient C_d using the well-known formula (1.3) or its analog

$$F = C_d S \rho_g (W_{rel})^2 / 2, \qquad (3.23)$$

where ρ_g is the gas (air) density and $S = \pi d^2/4$ is the mid-sectional area of the drops.

The results of calculating the drag coefficient in the spray flow according to the described method together with the data corresponding to formulas (1.4), (1.5), and (1.7) are shown in Figs. 3.10 and 3.11. Obviously, the hydrodynamic drag value 3, calculated from the experimental data, is significantly lower than that given by formulas (1.5) and (1.7), especially in the self-similar zone of the flow at $z > 300$ mm, where experimental value $C_d < 0.2$.

To explain noted anomaly of the drag coefficient, we put forward a number of following hypotheses:

1. Polydispersity of spray, which causes a difference in the motion of drops of different sizes and the appearance of singularities in the averaged, integral, motion of the dispersed phase
2. Deformation of the droplets in turbulent flow, in particular, the vibrations of their shape with a change in the particle size that is transverse to the flow
3. Macroscopic local inhomogeneity of the drop flow structure, which shows up as the group motion of the drops
4. Direct effect of velocity pulsations in a gas flow about a drop

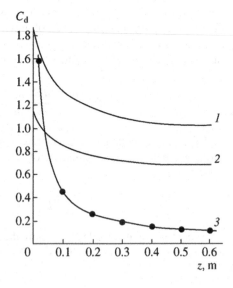

Fig. 3.10 Change in the drag coefficient by spray flow height; 1, 2, calculation by using (1.5) and (1.7), respectively; 3, according to the experiment data in a spray flow of a nozzle; $P_1 = 5$ atm [13]

Fig. 3.11 Dependence C_d(Re) obtained by calculation: 1, by Klyachko formula (1.5); 2, by Stokes formula (1.4); 3, the approximation $C_d = 2000/Re^2$ of the experiment data (*o*) for a nozzle spray [13]

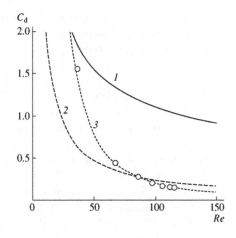

5. "Tightening" of Stokes regime of gas flow around the drop to the region of transient Reynolds numbers Re $\approx$ 1–100 due to high turbulence of gas flow incoming on drops
6. Crisis of drag which is well known for flow around a rigid sphere at Re $\sim 10^5$ and observed even at Re $\geq$ 50 because of a high turbulence of the gas flow

In the analysis of hypotheses on the basis of available experimental and literary data, the first two of them were rejected immediately as insolvent. Indeed, if a polydisperse set of drops is replaced by a monodisperse one with a mean size d_{32} and the same total mass and surface area of the drops, the mean drag coefficient may change by several tens of percent rather than by several times. Deformation, flattening of the drops in the direction of relative velocity of the phases, may raise but not decrease the drag force and coefficient [14].

Preference was initially given to the third hypothesis, according to which in a spray flow of a mechanical nozzle, the particles of a dispersed liquid move not separately but mainly in groups in the form of clusters, agglomerates, clots, and clouds [12]. These droplet groups occur during the decay of individual "filaments" and liquid films formed during the decay of a jet that flows from the nozzle [15]. The group movement of droplets differs in that within the clots the volume concentration of the liquid is higher and between them is lower than the average concentration in the volume containing a sufficient number of these clusters. This is what is meant by the term macroscopic local inhomogeneity of the structure of the drop stream, taking into account the fact that the dimensions of bunches are much larger than the sizes of the individual drops. In addition, when droplets of a group move together, the total gas-dynamic drag of droplets is less than the sum drag of the same droplets when they move separately. That explains the new phenomenon, which was firstly called not the crisis but by the anomaly of drag expressed in underestimated experimental values of C_d in Figs. 3.10 and 3.11 compared with values received by using Klyachko formula (1.5).

However, the third hypothesis also fails if one takes into consideration the fact that the anomaly (or crisis) of the drag arises at a certain distance from the nozzle (Fig. 3.10) where the concentration of the drops is appreciably lower than that in the top of the spray cone, where any anomaly is absent. In addition, the farther from the nozzle, the more turbulent pulsations of the gas flow must destroy the group movement of the droplets.

The fourth hypothesis, about the direct effect of gas velocity pulsations, also turns out to be untenable: due to this effect, the value C_d may increase, rather than decrease, and only by several percent.

The fifth hypothesis relies on the analogy with liquid flow through tube: in going from the laminar to the turbulent regime, the drag coefficient reaches a local minimum [10]. Using the concept of the laminar viscous surface layer that borders the turbulent boundary layer from the outside (the idea similar to that used in the theory of near-wall turbulence [9, 10]), we even succeeded in constructing a model accounting for the extension of the Stokes regime of the flow about a drop into the range of transitional Reynolds numbers Re = 50–120 (Figs. 3.11 and 3.12). But the construction of the model required a number of poorly substantiated assumptions to be made. Also, this model yet cannot answer the questions why the (pseudo)Stokes regime is absent near the cone top of a spray, where Reynolds numbers are smaller, and why the drag anomaly emerges only at distances $z > 100$ mm from the nozzle (Figs. 3.10 and 3.11).

Thus, at our disposal, the only sixth hypothesis seems the most priority, in accordance with which in a spray flow of a nozzle, the "early" crisis of drag for drops actually arises. It can take place even at Re ≈ 50, and not only at 10^5, and can be caused by a high degree of turbulence of the gaseous flow around drops. In this regard, let us pay attention to the following circumstances.

Fig. 3.12 The drag coefficient of a sphere as a function of the Reynolds number: 1, Rayleigh curve [9]; 2, on the Stokes theoretical formula (1.4), the circle outlines the region for the spray flow that is critical over Reynolds number; 3, decrease in C_d in the marked region in spray experiment [13]

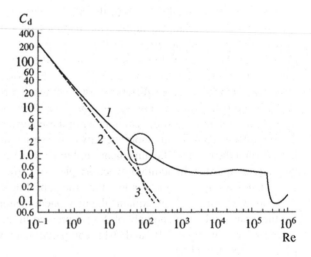

The dependence of the drag coefficient $C_d(\mathrm{Re})$ for a solid sphere on the Reynolds number in the range $10^{-2} < \mathrm{Re} < 10^6$ is well known. It was obtained by generalizing a large body of experimental data [9, 11, 14], and its plot (Fig. 3.12), taken from [9], is sometimes called the Rayleigh curve [16].

This curve can be subdivided into several portions. At $\mathrm{Re} < 1$, the flow is laminar and is described by the Stokes formula (1.4).

At $\mathrm{Re} \approx 20$, the laminar boundary layer in the stern region (a polar angle $\theta \leq 180°$) separates from sphere surface to form return vortex flow of the continuous phase [14, 16].

For $20 < \mathrm{Re} < 100$, the line of separation shifts upstream up to $\theta \sim 120°$, and the vortex behind the sphere grow up to 1.2 the sphere diameter. The drag coefficient exceeds the values from Stokes formula and can be approximated (up to $\mathrm{Re} = 400$–500) by Klyachko formula (1.5) or by the formula taken from [14, 16, 17].

$$C_d = 18.5/\mathrm{Re}^{3/5}. \tag{3.24}$$

At $\mathrm{Re} \approx 500$, the vortex is separating from the sphere by the flow and drifting downstream in the stern wake. At a certain point T downstream the line of separation, the transition to turbulent flow occurs in the stern wake behind the sphere [10].

At $500 < \mathrm{Re} < 10^5$, the flow about the sphere is usually called turbulent. However, it would be more appropriate to call it the mixed flow, since it is laminar upstream from the line of boundary layer separation ($\theta \approx 80°$) and turbulent downstream from the point T. As the Reynolds number grows, the point T moves upstream toward the spherical surface. When this point reaches the line of boundary layer separation ($\theta \approx 80°$), the flow throughout the separated layer becomes turbulent, and the drag coefficient in this wide region of Re remains almost unchanged ($C_d \approx \mathrm{const} \approx 0.5$).

At critical number $\mathrm{Re}_{cr} \approx 2.5 \times 10^5$, we come up against the well-known crisis of hydrodynamic drag for a sphere: the drag coefficient C_d decreases drastically down to 0.1–0.2 [9–11, 14]. The onset of the crisis is accompanied by the separation from the sphere of the laminar boundary layer, which becomes turbulent. Simultaneously, the line of separation shifts downstream toward $\theta = 120$–140°, that is, downstream from the transition point T at $\theta_T \approx 100°$. As this takes place, the flow about the sphere approaches the idealized flow pattern, and the pressure in the stern region of the sphere increases markedly, reducing the total hydrodynamic drag [10].

The aforesaid refers to the case when the flow about a sphere is initially laminar. It is known that "the turbulence of the incoming flow affects the crisis of drag. The higher turbulence, the smaller Re at which boundary layer turbulization occurs. As a result, the decrease in the drag coefficient starts at lesser Reynolds numbers (and is observed in a wider interval of Re)" [11]. It was reported in work [10] that, as the degree of turbulence $\varepsilon = w'_m/\langle w \rangle$ increases from 0.5% to 2.5% (where w'_m is the velocity pulsation amplitude and $\langle w \rangle$ is the averaged gas velocity), the critical Reynolds number Re_{cr} decreases from 2.70×10^5 to 1.25×10^5 (i.e., by half and not by three orders of magnitude!). In [10], it is also noted that "the crisis of drag may occur at Reynolds numbers that are considerably smaller than the critical value if the boundary layer is artificially turbulized."

It turned out that, when the turbulence of the flow is high ($\varepsilon \approx 30\%$ or higher, e.g., as in a spray flow produced by a nozzle [12]), the crisis of drag may occur at transitional Reynolds numbers Re > 50, which follows from the experiment [12].

The early crisis of hydrodynamic drag for drops in the spray of a nozzle explains a number of features of the two-phase flow that were detected in the experiment. These are the following:

1. Even at a distance of 1 m from the nozzle, the relative velocity of the phases is considerable, reaching 12 m/s (Fig. 3.9).
2. The momentum transferred from liquid to gas both in the free spray flow and in the spray apparatus is approximately half the initial momentum of the liquid jet.
3. The crisis arises at a certain distance from the nozzle rather than in the immediate vicinity of it. Therefore, two flow regions can be distinguished in the spray of a nozzle: (i) the active region, where the phases interact extensively, exchanging the momentum, and (ii) the self-similar region at distances $z > 300$ mm, where interaction between the phases is much weaker, their momentum fluxes are almost invariable, and the radial profiles of all the hydrodynamic characteristics are self-similar (see Sects. 3.1, 3.2, and 3.3 and [12]).

Numerical simulations of the two-phase flow in the spray of a nozzle in the framework of the two-dimensional model based on formulas (1.5) or (1.7) (i.e., without allowance for the early drag crisis) were invariably failed: the calculated gas velocities are much higher, and the liquid velocities are much lower than in experiments [12]. Conversely, when experimental values of C_d turning into account the early crisis were used in calculating the force of interfacial interaction, agreement between calculated and experimental velocities of the phases was greatly improved [18].

Thus, we conclude that highly turbulent flows like those in sprays produced by nozzles may exhibit early crisis of hydrodynamic drag for dispersed phase particles even at transitional Reynolds numbers such as several tens but not only at $\sim 10^5$, as was considered earlier.

3.5   Conclusions of Experimental Study of Spray Flow Hydrodynamics Created by a Nozzle

Analysis of the experimental results showed that:

1. The two-phase flow in a free spray of a mechanical nozzle is strongly turbulent.
2. The motion of gas in the spray flow has a close analogy with its motion in a flooded turbulent jet.

3. The two-phase spray flow is characterized by the presence of two zones: an active zone near a root of the stream and a self-similar zone starting at distances $z = 200\text{--}300$ mm from the nozzle.
4. In the first from these two zones, there is an intensive exchange of momentum between phases, and, probably, the most intense heat and mass transfer occurs.
5. In the self-similar zone, the intensity of the interphase transfer processes is much lower than in the active zone: the total pulse flux of each phase varies insignificantly, the radial distributions of the phase velocities, as well as the concentrations and specific fluxes of the liquid, have self-similar profiles, while the gas and liquid velocities in the spraying stream differ markedly from each other:
6. The value of the drop drag coefficient calculated from the experimental data in the self-similar zone turned out to be significantly lower (by four to seven times) compared with the corresponding values known from the literature, which is explained by the "early" (even at Re > 50) crisis of drag due to the high turbulence of a gas flow.

Some of the results presented in this chapter were previously published in articles [13, 19].

References

1. Nigmatulin, R. I. (1987). *Dynamics of multiphase media, part 1*. Moscow: Nauka. Hemisphere, New York, 1991.
2. Abramovich, G. N. (1960). *The theory of turbulent jets*. Moscow: Fizmatgiz.
3. Abramovich, G. N., et al. (1975). *Turbulent currents under the influence of bulk forces and non-self-similarity*. Moscow: Mashinostroyeniye.
4. Abramovich, G. N. (1970). On the influence of an admixture of solid particles or droplets on the structure of a turbulent gas jet. *DAN SSSR (Reports of AS USSR), 190*(5), 1052–1055.
5. Abramovich, G. N., et al. (1972). Turbulent jet with heavy impurities. *Izvestiya AN SSSR. Mekh. Zhidkosti i Gaza (Mechanics of Fluid and Gas)*, (5), 41–49.
6. Abramovich, G. N., & Girshovich, T. A. (1972). The initial part of a turbulent jet containing heavy impurities in a spiral stream. In *Investigations of two-phase, magneto-hydrodynamic and swirling turbulent jets* (Proceedings of the MAI, Moscow), No. 40, pp. 5–24.
7. Abramovich, G. N., & Girshovich, T. A. (1973). On the diffusion of heavy particles in turbulent flows. *DAN SSSR (Reports of AS USSR), 212*(3), 573–576.
8. Rychkov, D. A., & Shraiber, A. A. (1985). Axisymmetric polydisperse two-phase flow with coagulation and fragmentation of particles for an arbitrary fragment distribution by mass and velocity. *Izvestiya AN SSSR. Mekh. Zhidkosti i Gaza (Mechanics of Fluid and Gas)*, (3), 73–79.
9. Schlichting, H. (1955). *Boundary-layer theory*. New York: McGraw-Hill. Nauka, Moscow, 1969.
10. Loitsyanskii, G. G. (1978). *Mechanics of liquids and gases* (p. 736). Moscow: Nauka.
11. Landau, L. D., & Lifshits, E. M. (1986). *Course of theoretical physics* (Fluid mechanics) (Vol. 6). Moscow: Nauka. Pergamon, New York, 1987.
12. Simakov, N. N. (1987). Dissertation, Yaroslavl Polytechnic Inst., Yaroslavl.
13. Simakov, N. N. (2004). Crisis of Hydrodynamic Drag of Drops in the Two-Phase Turbulent Flow of a Spray Produced by a Mechanical Nozzle at Transition Reynolds Numbers. *Technical Physics, 49*, 188.

14. Nigmatulin, R. I. (1978). *Fundamentals of mechanics of heterogeneous media*. Moscow: Nauka.
15. Borodin, V. A., et al. (1967). *Spraying of liquids*. Moscow: Mashinostroenie.
16. Brounshtein, B. I., & Fishbein, G. A. (1977). *Hydrodynamics, mass and heat transfer in disperse systems*. Leningrad: Khimiya.
17. Bird, R. B., et al. (1960). *Transport phenomena*. New York: Wiley. Khimiya, Moscow, 1974.
18. Simakov, N. N. (2002). Numerical simulation of a two-phase flow in the spray stream produced by the nozzle. *Izvestiia Vysshykh Uchebnykh Zavedenii. Khimiya and KhimicheskayaTekhnologiya (News of universities. Chemistry and chemical technology), 45*(7), 125–129.
19. Simakov, N. N., & Simakov, A. N. J. (2005). Anomaly of gas drag force on liquid droplets in a turbulent two-phase flow produced by a mechanical jet sprayer at intermediate Reynolds numbers. *Applied Physics, 97*, 114901.

Chapter 4
Mathematical Modeling of Hydrodynamics of an Axisymmetric Two-Phase Flow Produced by a Nozzle

Now, two approaches are known for simulating a two-phase flow of sprayed liquid in a gas by using an injector. These are a method of mutually penetrating continua and the method of turbulent jets. Taking into account the flow peculiarities revealed in the analysis of the experimental results, it turned out that none of the abovementioned approaches taken separately can be used to model the spray flow. But it is possible to use their combination.

4.1 Combination of Known Phenomenological Approaches when Describing a Two-Phase Flow Created by Mechanical Injector

An analysis of the physical picture of a two-phase spray flow, based on experimental data, showed:

1. The average velocities of the phases are visibly different at all points of the flow.
2. There is a noticeable rarefaction of the gas in the root zone of the spray; by this it needs to take into account the equations describing the motion of the gas phase.
3. The formation of the spatial distribution of gas velocities is significantly influenced by turbulent momentum exchange (turbulent friction) inside the continuous phase.
4. The known analytical approximations for the density of the interphase interaction forces, in particular formulas (1.5) and (1.7) for the drag coefficient of dispersed liquid, are not always suitable for describing the hydrodynamics of the spray due to the occurrence of the crisis of drag already at the transit values of Reynolds number Re.

Thus, none of the two known phenomenological approaches to the modeling of hydrodynamics of a two-phase flow of spray, which were described in Chap. 1,

© Springer Nature Switzerland AG 2020

N. N. Simakov, *Liquid Spray from Nozzles*, Innovation and Discovery in Russian Science and Engineering, https://doi.org/10.1007/978-3-030-12446-5_4

specifically, the approach which is used, e.g., by R.I. Nigmatulin [1–4] and approach of G.N. Abramovich [5–10], taken separately, cannot be applied to the mathematical description of the spray flow of a mechanical nozzle, taking into account all its main regularities. The first approach does not take into account factors 1 and 2, and the second – factors 3 and 4 from the above.

At the same time, it is possible and very suitable to construct a two-dimensional mathematical model of an axisymmetric two-phase flow by using the combination of both above approaches, taking into account all the marked features of spray formed by a mechanical injector.

In such a combined mathematical description of the spray flow, the equations of continuity (1.10) and of momentum (1.11) are written separately for each phase, as in the continuum approximation [1, 3, 4]. In the right-hand parts of the time-averaged equations of the gas pulse, in addition to the terms describing the actions of gravity, pressure, and interfacial interaction, also the terms involving turbulent friction in the gas must be used analogy to turbulent jets [5, 6]. At the same time, viscous friction inside a gas, as will be shown below, can be neglected. The terms in the equations of phase pulses which describe the interphase interaction should take into account the feature noted above – the emergence of a resistance crisis in the flow past the drops with gas.

If it is necessary to take into account the forces of a static pressure in the gas, then in a set of the basic equations describing the two-phase flow, along with the equations of the continuity and momentum of phases, the energy equations must also be used. The kind and form of the latter can be different: separate equations for the kinetic and internal energy or the general equation for the total energy recorded for each phase separately or for the two-phase system as a whole [1, 3, 4]. In these equations, the work of the interphase interaction forces as well as the heat and mass transfer should be taken into account, if their effect on hydrodynamics is essential in the problem being solved.

In some cases, for example, when describing a two-phase flow in a self-similar zone of a free spray, the gradient of the static pressure in the gas can be neglected. In this case, there is no need to involve the energy equation, and the system of basic equations turns out to be simpler. Moreover, in simulating the hydrodynamics of a "cold" torch, only two basic equations are sufficient to describe the motion of an incompressible gas in a cylindrical coordinate system: the continuity equation and the equations for the axial component of the gas momentum, in each of which the porosity difference from unity can be neglected. The motion of the dispersed phase is described by three equations: the equation of continuity and two equations for axial and radial components of impulse. The equations listed can be written in conservative or nonconservative form [11, 12], in a nonstationary form with respect to the total values of the dependent variables or in a quasi-stationary form for time-averaged values of these quantities. In the latter case, additional terms appear in the equations, taking into account the pulsing components of the dependent variables.

For a vertically oriented downward two-phase flow, the initial system of basic hydrodynamic equations, corresponding to the simplified case, in the nonstationary nonconservative unfolded form has the following form:

$$\frac{\partial w_z}{\partial z} + \frac{1}{r} \cdot \frac{\partial (r \cdot w_r)}{\partial r} = 0, \tag{4.1}$$

$$\frac{\partial w_z}{\partial t} + w_z \cdot \frac{\partial w_z}{\partial z} + w_r \cdot \frac{\partial w_z}{\partial r} = g - \frac{\Phi_z}{\rho_g}, \tag{4.2}$$

$$\frac{\partial \alpha}{\partial t} + u_z \cdot \frac{\partial \alpha}{\partial z} + u_r \cdot \frac{\partial \alpha}{\partial r} + \alpha \cdot \frac{\partial u_z}{\partial z} + \frac{\alpha}{r} \cdot \frac{\partial (r \cdot u_r)}{\partial r} = 0, \tag{4.3}$$

$$\frac{\partial u_z}{\partial t} + u_z \cdot \frac{\partial u_z}{\partial z} + u_r \cdot \frac{\partial u_z}{\partial r} = g + \frac{\Phi_z}{\rho_l \cdot \alpha}, \tag{4.4}$$

$$\frac{\partial u_r}{\partial t} + u_z \cdot \frac{\partial u_r}{\partial z} + u_r \cdot \frac{\partial u_r}{\partial r} = \frac{\Phi_r}{\rho_l \cdot \alpha}, \tag{4.5}$$

where w_z, w_r, u_z, u_r are the velocity components of gas and liquid, respectively, α is the volume fraction of the liquid, g is the acceleration due to gravity, and Φ_z and Φ_r are the volume densities of the interfacial interaction forces.

Basic Eqs. (4.1–4.5) must be supplemented by expressions analogous to (1.6), which determine the dependence of the quantities Φ_z and Φ_r on other variables and will be specially considered below. Here we note that to solve the system of equations of hydrodynamics of the form (4.1–4.5), it is necessary to use the numerical methods and computers [11, 12].

These methods, in particular the large-particle method [4], include a specification of computational grid in a certain area of independent variable values; choice of a difference scheme approximating a system of differential equations and satisfying the stability condition of the calculation [11–13], which ensures the convergence of the difference solutions to the solution of the differential problem; specification of the boundary conditions; and initial values of the dependent variables on the computational domain.

Numerical methods for solving the system (4.1–4.5), which differ by the concrete form of the chosen difference scheme providing the stability conditions, have one common feature: when any of them is implementing on a computer, there are smoothing out effects of numerical diffusion and viscosity of artificial nature or due to approximation depending on the type of difference scheme [12]. In some cases, the influence of these numerical effects can substantially distort the received numerical solution. This influence needs to be taken into account and special measures should be taken to reduce it.

So, the application of the well-known Lax [12] scheme, which has a significant numerical viscosity on a large-scale grid, to the calculation of the spray flow hydrodynamics by the use of Eqs. (4.1–4.5), in which the interaction of phases was taken into account by formulas (1.3), (1.5), and (1.7) and turbulent friction in the gas was not taken into account at all, did not allow us to obtain results that could roughly be consistent with the experiment, precisely because of the numerical viscosity.

The results of our calculations on the same equations using other difference schemes, e.g., of Van Leer [14] and the large-particle method [4], proved more positive with regard to qualitative agreement with the experimental data. The absence of a quantitative agreement between the calculated and experimental results is expressed in the radial expansion of the calculated profiles of the gas velocity and, as a consequence, the double increasing of its flow through the cross section $z = 300$ mm in comparison with that found from the experiment and is explained by the incorrect calculation of the phase interaction by formulas (1.3), (1.5), and (1.7) and the fact that turbulent friction inside the gas was not taken into account in the calculation.

Thus, our first unsuccessful attempts to numerically simulate the hydrodynamics of a spray using a system of equations of the form (4.1–4.5) have shown that in order to accurately calculate the macroscopic structure of phase flows in the spray flow of a mechanical nozzle, i.e., structure of a locally averaged two-phase flow, it is necessary to take into account the turbulent friction of the gas and the features of the interfacial interaction.

4.2   Turbulent Friction of Gas in a Spray Flow

4.2.1   Simplification of Equations Describing Gas Flowing in the Self-Similar Zone of the Spray

To describe the axisymmetric flow of an incompressible gas, in the most general form, three basic equations of classical hydrodynamics can be used: the continuity Eq. (4.1) and two equations of gas motion for the axial (4.2) and radial component of its velocity. Taking into account all forces acting on the gas and smallness ($\sim 10^{-5}$) of the volume fraction of the liquid in the self-similar zone of the flow, the equations of motion with respect to the complete variables in cylindrical coordinates have the form

$$\frac{\partial w_z}{\partial t} + w_z \cdot \frac{\partial w_z}{\partial z} + w_r \cdot \frac{\partial w_z}{\partial r} = g - \frac{1}{\rho_g} \cdot \frac{\partial P}{\partial z} - \frac{\Phi_z}{\rho_g} + \nu \cdot \Delta w_z, \qquad (4.6)$$

$$\frac{\partial w_r}{\partial t} + w_z \cdot \frac{\partial w_r}{\partial z} + w_r \cdot \frac{\partial w_r}{\partial r} = -\frac{1}{\rho_g} \cdot \frac{\partial P}{\partial r} - \frac{\Phi_r}{\rho_g} + \nu \cdot \left(\Delta w_r - \frac{w_r}{r^2} \right), \qquad (4.7)$$

$$\Delta = \frac{\partial^2}{\partial r^2} + \frac{1}{r} \cdot \frac{\partial}{\partial r} + \frac{\partial^2}{\partial r^2}.$$

By dividing the variables on their characteristic values in the spray flow, namely, r and z (on the mean diameter $D_0 \sim 0.5$ m of the stream), w_r and w_z (on the typical value $u_0 \approx 20$ m/s of liquid velocity), t (on the characteristic time $t_0 = D_0/u_0 \sim 2.5 \cdot 10^{-2}$ s), and P (on the characteristic rarefaction $P_0 \approx 10$ Pa of the gas in the flow), we can proceed in Eqs. (4.1), (4.6), and (4.7) to dimensionless variables. If we retain for them the previous notations, then Eq. (4.1) does not change, and, for example, Eq. (4.6) takes the form

$$\frac{\partial w_z}{\partial t} + w_z \cdot \frac{\partial w_z}{\partial z} + w_r \cdot \frac{\partial w_z}{\partial r} = \frac{D_0}{u_0^2} \cdot \left[g - \frac{1}{\rho_g} \cdot \left(\frac{P_0}{D_0} \cdot \frac{\partial P}{\partial z} + \Phi_z \right) + \nu \cdot \frac{u_0}{D_0^2} \cdot \Delta w_z \right].$$

(4.8)

Using the above values of D_0, u_0, and P_0, as well as the gas density $\rho_g \approx 1.16$ kg/m^3, its kinematic viscosity $\nu \approx 1.55 \cdot 10^{-5}$ m^2/s, and the acceleration of gravity $g \approx 10$ m/s^2, we can present the *right-hand side* (RHS) of (4.8) in the form

$$\text{RHS} = 1.25 \cdot 10^{-3} \cdot \left[10 - 0.86 \cdot (20 \cdot \partial P/\partial z + \Phi_z) + 1.24 \cdot 10^{-3} \cdot \Delta w_z \right] \quad (4.9)$$

The maximum value of w_z, the values of its derivatives, and the value of the derivative $\partial P/\partial z$ in (4.8) and (4.9) are of the order of unity. An estimate of the interfacial interaction by formula (1.6) using experimental data for the points of the spray flow with the coordinates $r = 0$ and $z = 300$–500 mm gives $\Phi_z \approx -20$ Pa/m, the signs of $\partial P/\partial z$ and Φ_z are opposite, and the absolute value of the quantity Φ_z is high. Taking into account the remarks made, we establish that in order of magnitude, the RHS of (4.8) is $<10^{-2}$, and its value is 2 orders of magnitude smaller than the value (of order 1) of the left-hand side of (4.8); therefore we can practically neglect it.

In other words, in the self-similar zone of the spray flow, all forces whose action is described by the right-hand side of Eq. (4.6) can be neglected, i.e., first of all by the force of viscous friction and then forces of gravitation, pressure, and interfacial interaction, and we can assume that the movement of gas in this zone occurs by inertia.

From three equations of system, (4.1), (4.6), and (4.7), describing the motion of the gas, now only two equations are necessary – in terms of the number of independent variables w_z and w_r. In this case a simplified system of equations

$$\frac{\partial w_z}{\partial z} + \frac{1}{r} \cdot \frac{\partial (r \cdot w_r)}{\partial r} = 0, \qquad (4.10)$$

$$\frac{\partial w_z}{\partial t} + w_z \cdot \frac{\partial w_z}{\partial z} + w_r \cdot \frac{\partial w_z}{\partial r} = 0, \qquad (4.11)$$

written with respect to the components of the total velocity of the gas in the spray flow is completely analogous to a system describing by using the same variables the motion of an incompressible gas in an axisymmetric flooded turbulent jet [5].

Continuing the above analogy and using (4.10), we present the motion Eq. (4.11) to the form

$$\frac{\partial w_z}{\partial t} + \frac{\partial w_z^2}{\partial z} + \frac{1}{r} \cdot \frac{\partial (r \cdot w_z \cdot w_r)}{\partial r} = 0 \qquad (4.12)$$

Let us replace in (4.10) and (4.12) the real gas velocity by the sum of its averaged (in time) and pulsating components:

$$w = \bar{w} + w', \qquad (4.13)$$

after that we average the terms of the resulting equations in time, neglect in them a small term $\overline{(w_z')}^2 \ll \overline{(w_z)}^2$, and use the following notations for the stress of turbulent friction:

$$\tau = -\rho_g \cdot \overline{w_z' \cdot w_r'}, \qquad (4.14)$$

and instead of (4.10), we obtain exactly the same equation, but already with respect to the averaged components of the gas velocity. And instead of (4.12) in the quasi-stationary case, when $\partial \overline{w_z}/\partial t = 0$, we obtain equation

$$\overline{w_z} \cdot \frac{\partial \overline{w_z}}{\partial z} + \overline{w_r} \cdot \frac{\partial \overline{w_z}}{\partial r} = \frac{1}{\rho_g \cdot r} \cdot \frac{\partial (r \cdot \tau)}{\partial r}. \qquad (4.15)$$

To solve the system of Reynolds equations (4.10) and (4.15) relatively the averaged velocity components w_z and w_r, it is necessary to determine the form of the function:

$$\tau = \tau \left(r, \ z, \ \overline{w_z}, \ \overline{w_r} \right). \qquad (4.16)$$

To do this, by analogy with turbulent flooded jets, one could use for the turbulent friction the Prandtl formula [5].

$$\tau = -\rho_g \cdot L^2 \cdot \left(\frac{\partial \overline{w_z}}{\partial r} \right)^2, \quad L = k_1 \cdot Z = k_1 \cdot (z + Z_0), \qquad (4.17)$$

where $k_1 = $ const is the experimental constant. In this case, the solution of the system (4.10) and (4.15) reduces to Tolmin problem for an axisymmetric turbulent source [5, pp. 89–99]. However, a comparison of the theoretical velocity profile obtained by Tolmin for a turbulent fluid jet with experimental data for the gas flow in a spray (Fig. 3.5) indicates their disagreement: the Tolmin profile drops steeper.

This also applies to the new theory of free turbulence of Prandtl and Gertler [5, p.111], in which turbulent friction is described by formula

$$\tau = \rho_g \cdot \varepsilon \cdot \frac{\partial \overline{w_z}}{\partial r}, \tag{4.18}$$

(here $\varepsilon = $ const is the coefficient of the kinematic turbulent viscosity – the experimental constant) and Reichard theory of turbulent mixing [5, p. 129], since the theoretical profiles of the gas velocity in an axisymmetric flooded jet, obtained within the framework of these theories, practically do not differ from the Tolmin profile and, therefore, also do not agree with the experimental data shown in Fig. 5.5 for the spray flow of an injector.

For the further it is expedient to carry out a comparative analysis of certain aspects in the theories of turbulence of Prandtl and Reichard.

4.2.2   Comparative Analysis of Prandtl and Reichard Theories of Turbulence

The initial system of equations describing the turbulent flow of incompressible gas in an axisymmetric flooded jet is given by Eqs. (4.10) and (4.12) in both theories under consideration.

Further in the Prandtl theory, the variables are replaced on formula (4.13), and the equations are averaged over time, as a result of which the system of Eqs. (4.10) and (4.15) is obtained in the above-described way.

In contrast, in the Reichard theory, the variables are not replaced on formula (4.13) before averaging equations over time. And from (4.12) for quasi-stationary flow $\partial \overline{w_z}/\partial t = 0$, after averaging over time, instead (4.15) the equation follows:

$$\frac{\partial \overline{w_z^2}}{\partial z} + \frac{1}{r} \cdot \frac{\partial \left(r \cdot \overline{w_z \cdot w_r} \right)}{\partial r} = 0. \tag{4.19}$$

Then it is suggested that

$$\overline{w_z \cdot w_r} = -\Lambda(z) \cdot \frac{\partial \overline{w_z^2}}{\partial r}, \tag{4.20}$$

and from the self-similarity condition it follows that

$$\Lambda(z) = k_2 \cdot Z = k_2 \cdot (z + Z_0), \tag{4.21}$$

where $k_2 = $ const is the experimental constant.

Thus, the equation describing the change in momentum in Reichard theory takes the form

$$\frac{\partial \overline{w_z^2}}{\partial z} = \frac{1}{r} \cdot \frac{\partial}{\partial r}\left(r \cdot \Lambda \cdot \frac{\partial \overline{w_z^2}}{\partial r}\right). \tag{4.22}$$

Comparing (4.15) with (4.22), we note that in the last equation, instead of the averaged velocity value $\overline{w_z}$, another variable $\overline{w_z^2}$ proportional to the average value $\rho_g \cdot \overline{w_z^2}$ of the density of the axial flow of the gas pulse is used.

Starting from (4.13) and taking into account that when averaging $\overline{w_z'} = 0$, we obtain a relation connecting the indicated variables:

$$\overline{w_z^2} = (w_z)^2 + \overline{(w_z')}^2. \tag{4.23}$$

The pulsating component of the gas velocity almost everywhere, except for the boundary region of the jet, is much smaller than the averaged:

$$\overline{(w_z')}^2 \ll \overline{(w_z)}^2, \tag{4.24}$$

Therefore, in the most essential region of the jet, where condition (4.24) is satisfied and $\overline{w_z^2} \cong (w_z)^2$, instead of Eq. (4.19), we can write

$$\overline{w_z} \cdot \frac{\partial \overline{w_z}}{\partial z} + \frac{1}{2 \cdot r} \cdot \frac{\partial \left(r \cdot \overline{w_z \cdot w_r}\right)}{\partial r} = 0. \tag{4.25}$$

The last equation can be considered as an analog of Eq. (4.15) in the Prandtl theory, which we shall repeat here for convenience:

$$\overline{w_z} \cdot \frac{\partial \overline{w_z}}{\partial z} + \overline{w_r} \cdot \frac{\partial \overline{w_z}}{\partial r} = \frac{1}{\rho_g \cdot r} \cdot \frac{\partial (r \cdot \tau)}{\partial r}. \tag{4.26}$$

Comparing Eqs. (4.25) and (4.26), one can conclude that the combined effect of radial convective transfer of momentum and turbulent friction (turbulent "diffusion" of impulses), described in the Prandtl theory by the second and third term (4.26), respectively, in Reichard theory it is taken into account only by the second term of the Eq. (4.25).

If, in addition, we take into account that in a turbulent flooded jet, the quantities of derivatives $\partial \overline{w_z}/\partial z$ and $\partial \overline{w_z}/\partial r$ are of the same order and the radial component of the averaged velocity is much smaller than its axial component

$$\overline{w_r} \ll \overline{w_z}, \tag{4.27}$$

then convection of the momentum transversely the flow taken into account by the second term in (4.26) can be neglected, as is done in the Schlichting theory for the turbulent wake behind the body [5, p. 146]. In this case, instead of (4.26), we obtain

$$\overline{w_z} \cdot \frac{\partial \overline{w_z}}{\partial z} - \frac{1}{\rho_g \cdot r} \cdot \frac{\partial(r \cdot \tau)}{\partial r} = 0. \tag{4.28}$$

Taking into account the fact that the Prandtl and Reichard theories are in equal agreement with the experimental data for single-phase jets [5], from the comparison of (4.28) and (4.25), we obtain for the turbulent friction stress τ along with (4.14) another expression

$$\tau = -\frac{1}{2} \cdot \rho_g \cdot \overline{w_z \cdot w_r}. \tag{4.29}$$

In addition, taking into account Eqs. (4.17), (4.20), (4.21), (4.23), and (4.24), we can write the chain of equations

$$-\rho_g \cdot L^2 \cdot \left(\frac{\partial \overline{w_z}}{\partial r}\right)^2 = -\rho_g \cdot k_1^2 \cdot Z^2 \cdot \left(\frac{\partial \overline{w_z}}{\partial r}\right)^2 =$$

$$= \frac{1}{2} \cdot \rho_g \cdot \Lambda \cdot \frac{\partial \left(\frac{\partial \overline{w_z}}{\partial r}\right)^2}{\partial r} = \rho_g \cdot k_2 \cdot Z \cdot \overline{w_z} \cdot \frac{\partial \overline{w_z}}{\partial r} \tag{4.30}$$

and, in particular, equation

$$-k_1^2 \cdot Z \cdot \frac{\partial \overline{w_z}}{\partial r} = k_2 \cdot \overline{w_z}, \tag{4.31}$$

deciding which, we get formulas

$$\overline{w_z} = w_m(z) \cdot \exp\left(-\gamma \cdot \frac{r}{Z}\right), \quad \gamma = \frac{k_2}{k_1^2} = \frac{\Lambda \cdot Z}{L^2}, \tag{4.32}$$

Thus, within the framework of the accepted assumptions, the Prandtl and Reichard theories lead to the same result – the exponential profile of the main component of the gas velocity in the jet.

The function $w_m(z)$ in (4.32) describes the change in the gas velocity on the axis of the jet. According to the experimental data for the sprayed flow from the nozzle, its form is determined by formula (3.5). The dimensionless profile of the axial velocity of the gas in the spray flow, described by formula (4.32) for $\gamma = C_2 \cdot P_1^{1/2}$, $C_2 = \ln 2 / C_1 = 5.25$ (atm)$^{-1/2}$, is shown in Fig. 3.5 as dashed curve. Obviously, this profile is in satisfactory agreement with the experimental data for the spray of nozzle.

4.2.3   Feature of Turbulent Friction of Gas in a Spray of Nozzle

The difference between solution (4.32) and the profiles obtained in the framework of the Prandtl and Reichard theories, expressed in the difference between the dashed and solid curves in Fig. 3.5, seems to be explained by the peculiarity of the turbulent transfer of the momentum of gas in a spray flow of a nozzle compared to the same phenomenon in a flooded gas jet. This feature can be caused by the presence of a dispersed liquid phase in the gas stream and should appear in the structure of the function (4.16) describing turbulent friction. The latter consideration is of particular interest in connection with the mathematical description and numerical modeling of the spray pattern and requires more detailed consideration.

In what follows, for simplicity, the intime averaging sign is omitted. Taking (3.5) into account, formula (4.32) takes the form

$$w_z = \frac{\alpha}{Z} \cdot \exp\left(-\gamma \cdot \frac{r}{Z}\right), \quad Z = z + Z_0, \quad \gamma = C_2 \cdot P_1^{1/2}. \tag{4.33}$$

From Eqs. (4.22)–(4.24) and also (4.28) and (4.17), we obtain, respectively,

$$r \cdot \frac{\partial w_z^2}{\partial z} = \frac{\partial}{\partial r}\left(r \cdot \Lambda \cdot \frac{\partial w_z^2}{\partial r}\right), \tag{4.34}$$

$$r \cdot \frac{\partial w_z^2}{\partial z} = \frac{\partial}{\partial r}\left[-2 \cdot r \cdot L^2 \cdot \left(\frac{\partial w_z}{\partial r}\right)^2\right]. \tag{4.35}$$

Substituting w_z from (4.33) into (4.34) and (4.35), we integrate these equations with respect to r and obtain for their left and right sides, respectively:

$$\int_0^r r \cdot \frac{\partial w_z^2}{\partial z} \cdot dr = -r^2 \cdot \frac{\alpha^2}{Z^3} \cdot \exp\left(-2 \cdot \gamma \cdot \frac{r}{Z}\right), \tag{4.36}$$

$$r \cdot \Lambda \cdot \frac{\partial w_z^2}{\partial z} = -2 \cdot r \cdot \gamma \cdot \Lambda \cdot \frac{\alpha^2}{Z^3} \cdot \exp\left(-2 \cdot \gamma \cdot \frac{r}{Z}\right), \tag{4.37}$$

$$-2 \cdot r \cdot L^2 \cdot \frac{\partial w_z^2}{\partial z} = -2 \cdot \gamma^2 \cdot L^2 \cdot \frac{r}{Z} \frac{\alpha^2}{Z^3} \cdot \exp\left(-2 \cdot \gamma \cdot \frac{r}{Z}\right). \tag{4.38}$$

Comparing (4.36)–(4.38) with each other, we find

$$\Lambda = \frac{r}{2 \cdot \gamma}, \quad L^2 = \frac{r \cdot Z}{2 \cdot \gamma^2}, \quad \frac{\Lambda \cdot Z}{L^2} = \gamma. \tag{4.39}$$

Taking into account the last equality in (4.39), we note that the result (4.32) obtained by solving (4.31) is preserved. However, now the empirical constant γ within the framework of the above assumptions about the relation of convective transport and turbulent diffusion of the pulse turned out to be the same in both turbulence theories – Prandtl and Reichard – at the time as two different constants k_1 and k_2 were previously required. In addition, according to (4.39), the parameters Λ and L, characterizing the turbulent mixing of the gas, depend in our case on the radial coordinate r, whereas in the above theories, they were directly proportional to the axial coordinate according to (4.17) and (4.21).

Using the analytic solution (4.33) of the equation of the axial momentum of the gas (4.28) and by integrating the continuity Eq. (4.10), we can find the radial velocity component of the gas w_r. Indeed, multiplying both sides of (4.10) by r and integrating over this variable, we obtain

$$r \cdot w_r = - \int_0^r r \cdot \frac{\partial w_z}{\partial z} \cdot dr = \frac{\alpha}{\gamma^2} \cdot (t^2 + t + 1) \cdot e^{-t} + F(z), \quad t = \gamma \cdot r/Z. \tag{4.40}$$

An arbitrary function $F(z)$ can be found from the boundary condition on the axis of the spray flow:

$$w_r(0, z) = 0. \tag{4.41}$$

From (4.40) for small t, it is not difficult to obtain

$$r \cdot w_r = \alpha/\gamma^2 \cdot \left[1 + 1/2 \cdot t^2 - 2/3 \cdot t^3 + O(t^3) \right] + F(z). \tag{4.42}$$

Comparing (4.41) and (4.42), we conclude that

$$F(z) = -\alpha/\gamma^2 \tag{4.43}$$

$$w_r = \frac{\alpha}{\gamma^2} \cdot \frac{1}{r} \cdot \left[(t^2 + t + 1) \cdot e^{-t} - 1 \right] \tag{4.44}$$

For small r, when $t = \gamma \cdot r/Z \ll 1$, from (4.42), (4.43), and (4.33), the two successive approximations follow:

$$w_r \cong \frac{1}{2} \cdot \frac{r}{Z} \cdot w_z, \quad \frac{w_r}{w_z} \cong \frac{1}{2} \cdot \frac{r}{Z} \cdot \left(1 - \frac{1}{3} \cdot \gamma \cdot \frac{r}{Z} \right) \tag{4.45}$$

Note that the first of the results (4.45) can be obtained within the framework of the accepted assumptions in another way, namely, on the basis of our hypothesis (4.29) on turbulent friction. Indeed, if we replace the variables by using (4.13), then on the one hand, we receive

$$\overline{w_z \cdot w_r} = \overline{w_z} \cdot \overline{w_r} + \overline{w_z' \cdot w_r'}. \tag{4.46}$$

On the other hand, (4.14) and (4.29) give

$$\overline{w_z \cdot w_r} = 2 \cdot \overline{w_z' \cdot w_r'}, \tag{4.47}$$

from which it follows that

$$\overline{w_z} \cdot \overline{w_r} \cong \overline{w_z' \cdot w_r'}, \tag{4.48}$$

or taking into account Eqs. (4.14), (4.17), (4.39), and (4.33), we can get

$$\overline{w_r} = \frac{\overline{w_z' \cdot w_r'}}{\overline{w_z}} = \frac{L^2}{\overline{w_z}} \cdot \left(\frac{\partial \overline{w_z}}{\partial r}\right)^2 = \frac{1}{2} \cdot \frac{r}{Z} \cdot \overline{w_z}. \tag{4.49}$$

The above remark on the coincidence of the results (4.45) and (4.49) can be considered as an argument in favor of the validity of the above hypothesis (4.29) on the stress of turbulent friction.

In closing of the analysis of the turbulent transfer of a gas pulse in the spray flow, it is worth noting that, according to experimental data about the pressure receiver orientation corresponding to the extreme values of the micromanometer data (for the maximum of the total pressure and minimum of the static pressure), the conclusion was made differenced from (4.45) and (4.49), namely, it is that

$$\left(\frac{w_r}{w_z}\right)_{\exp} = tg\,[\varphi(r, z)] = \frac{r}{z}. \tag{4.50}$$

The difference in these results can apparently be explained by the complexity of the structure of the turbulence in the spray flow, which is difficult to take into account in the framework of the simplest assumptions in interpreting the experimental data. In particular, using the device used in measuring the gas velocity, including a pressure receiver and a micromanometer, a time-averaged value of the gas braking pressure was recorded, proportional to the average value of the square of the gas velocity:

$$\overline{w^2} = \left(\overline{w}\right)^2 + \overline{\left(w'\right)^2} = \left(\overline{w_z}\right)^2 + \overline{\left(w_z'\right)^2} + \overline{\left(w_r\right)^2} + \overline{\left(w'\right)_r^2}. \tag{4.51}$$

The relation (4.50) with allowance for (4.51), (4.24), and (4.49) corresponds to the expression

$$\left(\frac{w_r}{w_z}\right)_{exp} = \frac{\sqrt{\left(\overline{w_r}\right)^2 + \overline{\left(w'_r\right)^2}}}{\sqrt{\left(\overline{w_z}\right)^2 + \overline{\left(w'_z\right)^2}}} \cong \frac{\sqrt{\overline{\left(w'_r\right)^2}}}{\overline{w_z}}. \tag{4.52}$$

Hence it can be concluded that the root mean square value of transverse gas velocity fluctuations is defined by formula

$$< w'_r > = \sqrt{\overline{\left(w'_r\right)^2}} \cong tg\varphi \cdot \overline{w_z} = \frac{r}{z} \cdot \overline{w_z}. \tag{4.53}$$

However, reliable confirmation of this conclusion requires additional studies. At the same time, the conclusion about the approximate isocentricity (in center of the nozzle outlet) of the gas motion lines in the spray flow, made during the initial analysis of the experimental gas velocity data, is called into question.

Formula (4.53) makes it possible to estimate the degree (intensity [15]) of the turbulence of the gas flow $\varepsilon \cong w'/\overline{w} \cong < w'_r > /\overline{w} \cong tg\varphi$. Thus, for medium radii values, where the droplet flux density is maximum, the degree of turbulence is $\varepsilon \cong tg 16° \cong 0.3$ or 30%, i.e., quite high, and at the boundary of the spray flow, even twice more.

Thus, when averaged of the equations of gas motion on time in the right side of the equation for the axial component of the momentum, an additional term occurs taking into account the pulsation components of gas speed or the so-called apparent force of turbulent friction (4.14)–(4.15). In the spray flow created by a nozzle, the structure of turbulence and, accordingly, the structure of the mathematical expression for turbulent stress are somewhat different than in a turbulent flooded jet:

$$\tau = -\rho_g \cdot \frac{r \cdot Z}{2 \cdot \gamma^2} \cdot \left(\frac{\partial \overline{w_z}}{\partial r}\right)^2, \quad \gamma = C_2 \cdot \sqrt{P_1}. \tag{4.54}$$

For the investigated injector, the empirical constant $C_2 = 5.3 \ (atm)^{-1/2}$.

For the dispersed phase considered as pseudo-continuous compressible fluid having a density $\rho_1 \cdot \alpha$ when averaged on time of the Eqs. (4.3)–(4.5), the structure of their quasi-stationary approximations is also changed: the new members occur which are dependent on the averaged multiplying of the pulsation values, $\overline{\alpha' \cdot u'_r}$, $\overline{\left(u'_r\right)^2}$, etc.

Most of these terms can be neglected, as in the equations for the gas phase, in view of the smallness of the pulsation components of momentum compared with the mean values of the total variables. For the remaining, it is necessary to obtain the ratios between the products of pulsation and average values of the quantities, i.e., to specify the structure of the additional terms of equations, as it was made for the turbulent friction force in a gas. Note that this stage of the development of the mathematical model of the spray flow requires additional experimental data on the motion of droplets in the stream, which itself is a very difficult task, as can be seen from the above theoretical analysis of the gas phase motion in the spray flow.

At the same time, in order to complete the construction of a two-dimensional mathematical model of hydrodynamics of an axisymmetric spray flow created by a mechanical injector, it is necessary to determine the dependence of the interfacial interaction force on the averaged on time characteristics of turbulent two-phase flow, taking into account its difference from values calculated from well-known dependences, which, as was shown in the previous chapter, is due to the crisis of drag.

In this case, as it turned out, this difference is not due to the lack of an additional component of the interphase interaction force that arises when the expression for Φ is averaged over time and depends on the mean values of the products α' and w'_{rel} but, namely, due to the original structure of well-known expressions for quantity Φ type of (1.3), (1.5), and (1.7) which are not suitable for the spray flow of a mechanical injector. Note that in the extreme case, one can use the values of the drag coefficient obtained in the experiment.

The questions posed at the end of this subsection are not trivial and require serious research. Therefore, it seems advisable to refer also to simpler – albeit less general – models of the hydrodynamics of the spray flow of a mechanical injector.

4.3   Conclusions on the Initial Modeling of the Spray Flow

1. In the mathematical modeling of the hydrodynamics of the spray flow of a mechanical nozzle, taking into account all its main features and, first of all, the difference in phase velocities and the developed turbulence of a two-phase flow, it is necessary to use a combination of two known approaches to the description of multiphase flows: the model of turbulent two-phase jet and the model of two-speed interpenetrating continua.

2. In the equations of classical hydrodynamics of the continuity, the amount of motion of each phase and energy, it is necessary to use the averaging of these equations by time to go from the full values of the time-dependent variables to their averaged components, the values of which, as a rule, are determined experimentally. With this transformation of the equations, new additional terms appear in them, depending on the time-averaged multiplications of the pulsations components of the variables. In view of their smallness, some of these new terms of the equations can be neglected. The mathematical structure (i.e., the specific form of the dependences from the averaged components of variables) of the remaining terms describing turbulent diffusion and friction must be solved on the basis of an analysis of the experimental data, which is not a trivial task. But it is necessary for a successful numerical solution of the averaged equations of Reynolds.

3. The structure of the stress of turbulent friction of gas in the spray flow of a mechanical injector differs from that in a single-phase turbulent flooded gas jet: the length of the mixing path L or Λ depends not only on the axial but also the radial coordinate of the flow (4.39). Apparently, this is due to the nature of the stream, in particular, the presence of a dispersed phase in it.

References

1. Rychkov, A. D., & Shraiber, A. A. (1985). Axisymmetric polydisperse two-phase flow with coagulation and fragmentation of particles for an arbitrary fragment distribution by mass and velocity. *Izvestiya AN SSSR. Mekh. Zhidkosti i Gaza (Mechanism of Fluid and Gas)*, (3), 73–79.
2. Sou, S. (1971). *Hydrodynamics of multiphase systems* (Russian Transl.). (Mir, Moscow).
3. Nigmatulin, R. I. (1978). *Fundamentals of mechanics of heterogeneous media*. Moscow: Nauka.
4. Belotserkovsky, O. M., & Davydov, Y. M. (1982). *The method of large particles in gas dynamics*. Moscow: Nauka.
5. Abramovich, G. N. (1960). *The theory of turbulent jets*. Moscow: Fizmatgiz.
6. Abramovich, G. N., et al. (1975). *Turbulent currents under the influence of bulk forces and non-self-similarity*. Moscow: Mashinostroyeniye.
7. Abramovich, G. N. (1970). On the influence of an admixture of solid particles or droplets on the structure of a turbulent gas jet. *DAN SSSR (Reports of AS USSR), 190*(5), 1052–1055.
8. Abramovich, G. N., et al. (1972). Turbulent jet with heavy impurities. *Izvestiya AN SSSR. Mekh. Zhidkosti i Gaza (Mechanism of Fluid and Gas)*, (5), 41–49.
9. Abramovich, G. N., & Girshovich, T. A. (1972). The initial part of a turbulent jet containing heavy impurities in a spiral stream. In *Investigations of two-phase, magneto-hydrodynamic and swirling turbulent jets* (Proceedings of the MAI, Moscow), No. 40, pp. 5–24.
10. Abramovich, G. N., & Girshovich, T. A. (1973). On the diffusion of heavy particles in turbulent flows. *DAN SSSR (Reports of AS USSR), 212*(3), 573–576.
11. Potter, D. (1975). *Computational methods in physics* (Russian Transl.). (Mir, Moscow).
12. Roach, P. (1980). *Computational hydrodynamics* (Russian Transl.). (Mir, Moscow).
13. Dyachenko, V. F. (1977). *Osnovnyye ponyatiya vychislitel'noy matematiki (basic concepts of computational mathematics)*. Moscow: Nauka.
14. Van Leer, B. (1969). Stabilization of difference schemes for the equation of inviscid transfer problems in thick layers. *Journal of Computational Physics, 3*, 291–306.
15. Loitsyansky, G. G. (1978). *Mechanics of fluid and gas*. Moscow: Nauka.

Chapter 5
Testing Hypotheses About Reasons of Anomaly-Crisis of Particles Drag

To determine the cause of the anomaly (crisis) of drag for droplets in the spray flow, some hypotheses were tested on the effect of the group movement of droplets and the circulation motion of the liquid in them, the influence of geometry, and the strong turbulence of a gas flow around a droplet.

5.1 Experimental Confirmation of the Early Crisis of Drag for a Single Ball

The new phenomenon of drag early crisis initially observed for droplets in a two-phase turbulent flow at transitional Reynolds numbers (Re = 10–100) was experimentally found as well in the case when a gas flowed about a single hard sphere. With using a specially designed torsion balance, the drag force was measured for a foam plastic sphere 3 mm in diameter placed in a ventilator-produced turbulent jet. It is found that for the drag anomaly to arise, the turbulence of flow about the sphere must be sufficiently high.

5.1.1 Anomaly of Hydrodynamic Drag of Drops in a Spray of a Nozzle

Known methods of analyzing and calculating heat and mass exchange in two-phase system of droplets and a gas rely upon hydrodynamic equations allowing for interfacial interaction. However, existing knowledge and insight into the mechanism of interphase interaction in a two-phase flow resulting from liquid spraying certainly do not now suffice.

© Springer Nature Switzerland AG 2020
N. N. Simakov, *Liquid Spray from Nozzles*, Innovation and Discovery in Russian
Science and Engineering, https://doi.org/10.1007/978-3-030-12446-5_5

Hydrodynamic force F acting on spherical particle (e.g., on droplet) in a gas flow is usually defined by the formula (1.3) which is equivalent to formula (3.23). The drag coefficient depends on Reynolds number Re, characterizing a gas flow about a particle. For the case of Stokes laminar flow about particle (at Re < 0.1 ≪ 1), drag coefficient C_d can be found from the well-known and experimentally proven Stokes formula (1.4).

Reliable and sufficiently accurate theoretical formulas (such as formula (1.4)) adequately describing a laminar (a fortiori turbulent) flow about a particle at Re > 1 are lacking in the literature. In these cases, empiric data are invoked in practice. For example, having generalized a large body of experimental data obtained at Re > 1 for the steady fall of a sphere in a quiescent fluid and for a laminar flow around it in a duct and using the formula (1.4), Rayleigh presented for dependence C_d(Re) the standard unifying curve, which is often called the Rayleigh curve [1–3].

Portions of this curve can be approximated by analytical expressions, for example, by the Klyachko formula (1.5), which is applied in the transitional range 2 < Re < 700 for both the steady and accelerated motion of a spherical particle in a continuous medium [1, 4–6]. Function (1.5) is shown in Fig. 3.11 by a continuous curve *1*.

The motion conditions for a spherical particle in a continuous medium may differ noticeably from those under which the Rayleigh curve was derived (steady fluid laminar flow about a sphere with a uniform velocity field). Then, experimental values of C_d may tremendously ("by several thousand percent") diverge from the standard curve [1, 7]. Such discrepancies (anomalies) may be associated with the not spherical shape, rough surface, and accelerated motion of the particle, as well as with its rotation about the own axis. However, the primary reason for the discrepancies is the turbulence of the incident flow [1, 7].

The obvious feature of the Rayleigh standard curve is the abrupt (four- to fivefold) decrease in C_d at Re = $(2–3)\cdot10^5$, which is called the crisis of drag [1–3, 7, 8]. This phenomenon is explained by "improving" (streamlining) the flow near the sphere when the laminar boundary layer separates from its surface, becoming turbulent, and the line of separation shifts downstream [8].

It is known that the crisis of drag strongly depends on the degree or intensity of turbulence of the incident flow

$$\varepsilon = \frac{\sqrt{(w')^2}}{W}$$

which is the ratio of root mean square (rms) value of flow velocity pulsation to the time-averaged value W of its velocity. With increasing ε, the crisis comes "earlier," i.e., at lower values of Re [3, 7, 8]. For example, when ε for a gas flow in the aerodynamic tube increased from 0.5% to 2.5%, critical Reynolds number Re_c (corresponding to $C_d = 0.3$ for a sphere) decreased by more than twofold: from $2.70\cdot10^5$ to $1.25\cdot10^5$ [8].

The turbulence relative intensity is sometimes defined as

$$\varepsilon_1 = \frac{\sqrt{\overline{(w')^2}}}{V_r}$$

where $V_r = \bar{W} - U$ is the time-averaged fluid flow velocity relative to particle [7]. Clearly that, $\varepsilon = \varepsilon_1$ for a flow about a stationary ($U = 0$) sphere, $\varepsilon_1 > \varepsilon$ when the sphere moves in the same direction as the fluid, and $\varepsilon_1 < \varepsilon$ when speeds of sphere and fluid have the opposite directions. It was reported in work [7] that the crisis of drag usually observed for a sphere at Re $\sim 10^5$ shifts from $Re_c = 2200$ to $Re_c = 400$ with turbulence ε_1 varying from 14% to 40%.

Note that, on the standard curve C_d(Re), the crisis of drag shows up as a sharp (in several fold) decrease in C_d at Re $> Re_c$. Moreover, in a number of experiments, C_d recovered gradually to a previous standard value if Re increased to values exceeding Re_c by several times [7]. One can therefore conclude that the discovered crisis is of a local character, i.e., being observed in different parts of the wide interval $400 < Re < 10^6$. The following analytical dependence appropriate to experimental data was derived in [7]:

$$Re_c \cdot (\varepsilon_1)^2 \approx const = 45. \tag{5.1}$$

This formula relates a decrease in Re_c with an increase in ε_1.

Remarkably, the authors of [7] thought that the crisis of drag is hardly possible in the interval $1 < Re < 100$, since turbulence intensity ε_1 here would have to exceed 40%.

However, the anomaly (or early crisis) of drag for drops in a spray of a nozzle has been recently discovered in this Re range ($1 < Re < 100$) too [9–11]. Empty circles in Fig. 3.11 are data points for water drops in a spray of a nozzle [9], which are well approximated by the function

$$C_d = 2000/Re^2, \tag{5.2}$$

graph of which shown by dotted line in Fig. 3.11. It is seen from Fig. 3.11 that, as the drops move away from the nozzle (Re increases), their hydrodynamic drag diminishes from normal values determined by Klyachko formula (1.5) to anomalously low values that are close to those given by Stokes formula (1.4). In this case, the anomaly (crisis) also shows up in that, in the range $50 < Re < 130$, the drag coefficient, which found experimentally for drops in the spray of a nozzle, is four to seven times lower than that corresponding to the standard curve.

The fact is very remarkable [11, 12] that calculations by formula (5.2) for an axisymmetric nozzle spray using the 2D model with allowance for the anomaly (crisis) agree with the experiment fairly well (Fig. 5.1). However, results obtained for a nozzle spray without accounting the drag anomaly by Klyachko formula (1.5) disagree with experimental data (Fig. 5.2).

Fig. 5.1 Calculated (curves) vs. experimental (symbols) profiles of gas axial velocity W_z (circles) and liquid drop velocity U_z (squares) in the cross section of the nozzle spray at distance $z = 300$ mm from the nozzle [13]

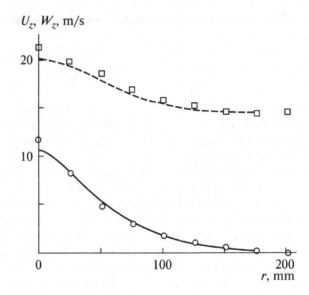

Fig. 5.2 Variation of gas flow relative velocity $W_{rel} = W_z(0, z)/U_0$ with distance along the axis of the flow (W_z is the gas flow axial velocity, and $U_0 = 26$ m/s is the initial velocity of the liquid drops). Squares, Miller experimental data [14]; continuous curve, calculation by the 2D model of the nozzle spray with allowance for the crisis of drag; and dashed curve, the same without allowance for the crisis of drag [13]

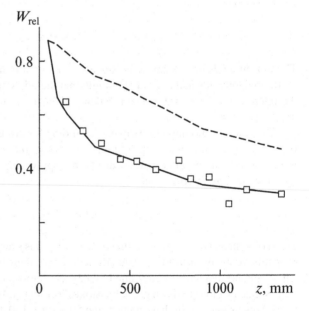

To explain the new drag anomaly in the range $1 < Re < 200$, six hypotheses were put forward initially [9–11]. Later only three of them remained which were to be worth looking at: (i) collective motion of drops, (ii) water circulation inside the drops (notwithstanding that their size is small, $d \sim 10^{-4}$ m and that the viscosity of water is much higher than that of air), and (iii) the early (already at $Re > 40$) crisis of drag that is similar to that observed in the case of a flow about a hard sphere at $Re \sim 10^5$.

The second and especially the third hypotheses assume that a high initial turbulence of gaseous flow ($\varepsilon > 30\%$ [9–11]) is critical for the anomaly (crisis) to occur, since $Re_1 = WD\rho/\mu \sim 10^5$ is high (W and D are the velocity and diameter of the flow, respectively).

The second hypothesis is supported by the well-known fact that the liquid circulation inside drops (which may be enhanced by a high turbulence of the gas flow about them) reduces their drag [6]. The third one stems from the similarity of the new anomaly to the classical crisis of drag for a hard sphere discovered at $Re \sim 10^5$. Since the new anomaly arises at much smaller Reynolds numbers ($50 < Re < 130$), it was called *the early crisis*.

To check each of the three hypotheses, it was necessary to make a new experiment in which a highly turbulent gas flows about a single hard spherical particle in the same range of transition Reynolds numbers ($Re = 10$–100).

5.1.2 Description and Experimental Results of Measuring the Drag Coefficient of a Single Sphere in a Turbulent Stream

Such an experiment has been conducted with the aim of verifying the crisis of drag for spherical particles at small (transitional) Reynolds numbers, $Re \sim 100$, and clarifying the real mechanism underlying this phenomenon. The measuring value was a drag coefficient C_d of a hard (made of foam plastic) sphere $d = 3$ mm in diameter placed in a highly turbulent air jet with initial diameter $D = 40$ cm produced by a household fan.

First of all, it was necessary to provide interaction conditions for a spherical particle and a turbulent gas flow similar to the conditions in the experiment with water drops in a spray of a nozzle [9]. These conditions, as was noted above, are determined by two Reynolds numbers. One, $Re = Vd\rho/\mu$, characterizes gas flow conditions around a particle and depends on diameter d of the particle and its velocity relative to the gas, $V = |\mathbf{U} - \mathbf{W}|$. The other Reynolds number, $Re_1 = WD\rho/\mu$, characterizes the incident gas flow as a whole and depends on gas flow velocity W and diameter D. In the spray of a mechanical nozzle, these numbers, $40 < Re < 130$ and $Re_1 \sim 10^5$, differ by roughly three orders of magnitude [9, 10]. Note that the conditions of flowing around the particle corresponded to the laminar-to-turbulent transition regime, $1 < Re < 500$, according to the adopted classification [3, 8]. At the same time, the flow incident on the particle was highly turbulent, $Re_1 \gg 2000$.

Conditions similar to those described above ($120 < Re < 530$, $Re_1 \approx 0.5 \cdot 10^5$) were provided by hanging a single foam sphere on a special-purpose torsion balance and placing it in a turbulent air jet produced by a fan.

The design of the torsion balance and the measuring scheme are shown in Fig. 5.3. Polymer filament *3* (in the given case, a fishing line $d_1 = 0.34$ mm in

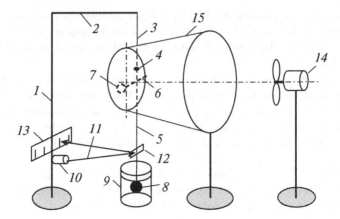

Fig. 5.3 Torsion balance (members *1–13*) for measuring the hydrodynamic drag force of sphere *7* in the air flow [13]

diameter) $L_1 = 500$ mm long is attached to horizontal rod *2* overhanging from vertical support *1*. The filament serves as an elastic member of the balance.

Metallic (guitar) string *5* $L_2 = 950$ mm in length and $d_2 = 0.25$ mm in diameter is rigidly fixed to the lower end of the filament at point *4*. Metallic balance arm *6* made of bronze wire $d_3 = 0.24$ mm in diameter (the length of either arm is $L_3 = 29$ mm) is soldered to the string perpendicularly to it 50 mm below point *4*, at which the string is attached to the filament. Light foam sphere *7* $d = 3$ mm in diameter is glued to the end of one arm. Massive ($m = 112.7$ g) metallic sphere *8* with diameter $d_4 = 30.5$ mm, which imparts considerable inertia moment $I = 1.05 \times 10^{-5}$ kg m^2 and natural period of vibrations $T \approx 16$ s to the torsion balance, was fixed to the lower end of the string coaxially with it. Sphere *8* is immersed in glycerol-filled vessel *9* to provide heavy decay of the natural vibrations of the balance. Small light source *10* (laser pointer) is fixed to support *1*. A thin light beam *11* from source *10* is incident small ($S = 5 \times 5$ mm^2) planar mirror *12* rigidly fixed to the lower part of the string. The beam reflected from the mirror is incident on flat scale *13* also fixed to support *1*.

The torsion balance operates as follows. A gas flow about sphere *7* generates hydrodynamic force **F**, which acts on the sphere and produces moment **M**. This moment twists elastic filament *4* and turns the balance arms together with the string and mirror. The angle of rotation of the arms, which is proportional to torque **M**, is determined from the shift of the light spot on scale *13*.

Thus, the torsion balance as a device measuring hydrodynamic force **F** acting on sphere *7* comprises members *1–13* (Fig. 5.3). A turbulent air jet flowing about sphere *7* was produced by fan *14*. In another design of the experiment, the sensitive element of the balance (arms *6* with sphere *7*) was placed near the end of confuser *15*, which was made of paperboard in the form of the lateral surface of a truncated cone and placed coaxially with an air jet generated by fan *14*.

Fig. 5.4 Inverse value $1/W$ to the air velocity versus axial coordinate z of the turbulent jet (distance from the fan); squares, data points; straight line, their linear approximation [13]

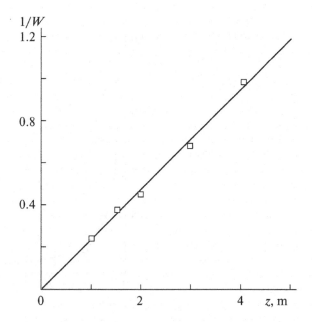

At the first stage of the experiment, the torsion balance was placed directly in the air jet produced by the fan in the absence of the confuser. The arms were located near the axis of the jet. The velocity of the air flow about the sphere was controlled by varying either a distance from the sphere to the fan or the rotation velocity of the fan blades. Near the sphere, the air velocity was measured with an Xplorer 1 impeller anemometer. First, the air velocity was measured at some "point" of the flow, and then the sphere on the balance arm was placed at the same point instead of the anemometer. The air velocity was measured at a given point of the flow for a given rotation velocity of the fan 25 (sometimes 50) times. Digital readings of the anemometer were taken visually and written into a register book manually. From these measurements, the mean value of velocity W and its random error were determined. The latter varied from 2% to 15% in different measuring series, being equal to 5–7% in most cases.

Figure 5.4 shows the variation of the measured air velocity along the axis of the turbulent jet. Axial coordinate z of the jet reckoned from the fan is plotted on the abscissa axis, and the inverse value W^{-1} of air velocity at a given point of the flow is plotted on the vertical axis. The z dependence of W^{-1} is seen to be linear, which is typical of turbulent jets [3, 8] and to some extent confirms the reliability of the data for the air velocity in the jet measured with the given anemometer.

Figure 5.5 plots hydrodynamic drag coefficient C_d for the small sphere attached on the balance arm in the free air jet versus Reynolds number Re. Data points for the crisis of drag in the case of droplets in a spray of a nozzle are also shown for comparison. Note once again that the new interval of the crisis, $120 < \text{Re} < 530$, is a continuation (with a slight overlap) of the interval $40 < \text{Re} < 130$ found earlier, in which the anomaly of drag for droplets was discovered. Data points marked by

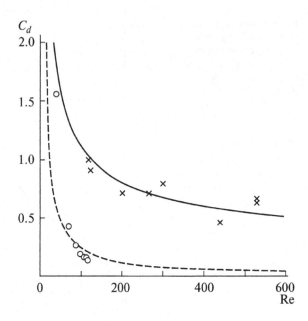

Fig. 5.5 Drag coefficient C_d of the spherical particle versus the Reynolds number. Crosses denote experimental data for the single sphere in the free air jet, solid line, calculation by formula (1.5); circles, experimental data for the spray of a nozzle; dash line, calculation by formula (1.4) [13]

crosses in Fig. 5.5 are placed near at the continuous curve (with account to a measurement error) corresponding to Klyachko formula (1.5). That is, contrary to expectation, the crisis of drag for the single sphere was not observed in a rather wide range of Reynolds numbers, $120 < Re < 530$, in spite of the high turbulence of the free air jet ($Re_1 \sim 0.5 \cdot 10^5 \gg 2000$).

Thus, the assumption that the crisis of drag in the case of drops is due to a high turbulence of the gas flow itself (the assumption underlying the second and third hypotheses) has not been confirmed. Conversely, Fig. 5.5 suggests that, even if a highly turbulent jet ($Re_1 \sim 0.5 \cdot 10^5$) flows about the sphere, the mean values of hydrodynamic force F and, accordingly, drag coefficient C_d may remain nearly the same as in the standard curve corresponding to a laminar flow. Such a conclusion is very important for simulating the motion both of a single sphere and of a two-phase disperse media.

Reflecting on why the experimental data did not meet the expectations to discover an early crisis for the case of a free jet flowing around one sphere, the author recalled the features of a two-phase flow in spray of a nozzle and an unusual picture of flowing of a gas around droplets and came to the following reasoning.

Drops resulting from the disintegration of a swirling liquid jet flowing from a swirl-type spray nozzle are scattering within a cone the apex angle of which is called the root angle of a spray of a nozzle. In our case, this angle equals 65° [9]. Moving drops entrain the surrounding gas (air), generating a turbulent jet. The gas moves like drops: it expands within the same cone in the form of a nonuniform flow with velocity W slower than velocity U of the drops ($W < U$). It can be said that each droplet in the stationary laboratory frame of reference which related to a nozzle moves in a gas flow that conically expands (as in a diffuser) and slows down, so a droplet outruns it. It was found experimentally that the relative velocity of the phases, $V = |U - W|$, in the free spray of a nozzle first increases [9, 10]. At the axis of the spray, it does not decrease down to 1 m from the nozzle (farther from the nozzle, measurements were not made). Now, mentally passing to the frame of reference related to a droplet, we will "see" an accelerating gas flow about a quiescent droplet, as it were, for example, in a confuser.

Thus, two possible patterns of the nonuniform gas flow relative the droplet can be realized in the above frames of reference: (i) the droplet outstrips the gas flow which is decelerating (as in a diffuser) and (ii) the droplets lags behind from the gas flow which is accelerating (as in a confuser).

Based on experimental data for the spray of a nozzle [9], one can assume that the specific flow conditions in these two cases can prevent the separation of the gas flow from the surface of the drop (or "shift" it to higher values of a relative phase velocity V). In this case, the flow conditions around the drop may be improved, and its drag may decrease, thereby producing the effect of C_d anomaly in the transitional range of Reynolds numbers, $1 < \text{Re} < 500$. In case (ii), this assumption seems to be even more comprehensible, if not obvious. In fact, a gas incident on a spherical (may, quiescent) particle in the form of a conically convergent flow which is accelerating reaches the trailing edge of the sphere faster than a flow with initially uniform velocity field with parallel flow lines. This must shift the arisen separation line downstream. One may expect in this case that the flow will become more streamlined (flow conditions will be improved) and C_d will decrease.

On the other hand, one can merely suppose that the value of the Reynolds number, $\text{Re}_1 = WD\rho/\mu \sim 0.5 \cdot 10^5$, and also the value of velocity pulsation amplitude w' turn out to be insufficient for the crisis of drag to occur for the sphere in the free jet from the fan. It is known that local intensity of turbulence ε varies along the radius of the jet (or of a flow in a tube), first increases and then decreases [1, 8, 9]. At the jet axis, the value Re_1 varies little on along jet length L reckoned from the pole, since characteristic diameter D of a turbulent jet increases in direct proportion to L and characteristic velocity W decreases in inverse proportion to L [3, 8]. However, in a tapered tube (confuser), Re_1, along with velocity pulsation amplitude w', may increase in inverse proportion to tube diameter D, with the gas flow rate (which is directly proportional to the product WD^2) remaining the same.

After the described above experiment with a free turbulent flow around a single sphere attached on a torsion balance, we conducted one more experiment to check the hypothesis that the nonuniform velocity field of a flow incident on the sphere influences the crisis of drag. This experiment differed from the previous one in that confuser *15* with the torsion balance at the exit was placed on the path of the air jet from fan *14* coaxially with the jet (Fig. 5.3). The sensitive member of the balance (arms *6* with foam sphere *7* on one arm) was placed inside the confuser near its axis and the outlet cross section. The confuser was made of paperboard in the form of the lateral surface of a truncated cone. The diameter of its inlet opening (the larger base of the cone) was 420 mm, that of the outlet opening, 135 mm (i.e., almost thrice as small), and the length of the generatrix, 470 mm. The geometry of the confuser was chosen so that the generatrix and the axis of the cone made an angle of 17°. The same

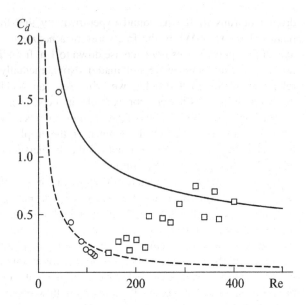

Fig. 5.6 Drag coefficient C_d of the spherical particle vs. the Reynolds number. Squares denote experimental data for an air jet flowing about a single sphere in the confuser; the other designations, as in Fig. 5.5 [13]

angle was made by the boundary of the spray of a nozzle in its self-similarity zone (i.e., at distance $z > 300$ mm from the nozzle) with the spray axis in the experiment [9]. Under these conditions, Re_1 increased almost threefold, having approached values which taken place inside the spray at experiment conditions [9].

In Fig. 5.6, experimental data for the hydrodynamic drag of the sphere about which an air jet flows in the confuser are shown by squares. Circles in this figure are experimental data on the early crisis of drag for drops in the spray of a nozzle. It is seen that, first, the drag anomaly is observed not only for drops in the spray but also for the flow about the single sphere in the confuser. Second, the interval $140 < Re < 300$, where the early crisis of drag for the single sphere was discovered, is a continuation of the interval $40 < Re < 130$, where the same crisis was discovered for liquid drops [9, 10]. This suggests the similarity of the crises.

The early crisis shows up most vividly in the interval $50 < Re < 200$, where measured drag coefficient C_d of the sphere was four to seven times smaller than the published data, for example, than those approximating experimental data by the Rayleigh curve in the transitional interval $2 < Re < 400$ accurate to 10% using Klyachko formula (1.5) [6]. Note that the interval $50 < Re < 200$ is rather wide: the relative velocity of the phases in it varies fourfold. Within the next interval, $200 < Re < 400$, drag coefficient C_d for the sphere gradually grows, approaching standard values in the Rayleigh curve, as follows from Fig. 5.6.

Which of the two factors, nonuniformity of the velocity field in the incoming flow or an additional increase in Re_1 and w' in it, is main to the occurrence of the crisis of drag for a sphere in the confuser was still unclear.

5.1.3 Conclusions About the Early Crisis of Particle Drag

Thus, the anomaly or early crisis of drag for drops in a spray of a nozzle discovered and described in [9, 10] has found an additional confirmation in experiments with the gas flow around the single hard sphere in the confuser. The results of both experiments obtained by different methods for radically different objects are in essence similar, indicating the general character of the new phenomenon.

As in the case of Reynolds numbers in the range $400 < \text{Re}_c < 2200$, the necessary condition for the early crisis of drag in the case of a gas flow about a single sphere is a high turbulence ($\text{Re}_1 \sim 10^5$) of the flow with turbulence intensity $\varepsilon > 30\%$. The reason for this phenomenon is an improved flow pattern around the sphere (more streamlined flow) under these conditions. This makes it possible to prevent the separation of the boundary layer from the surface of the sphere or to shift a separation to higher values of V. If separation still takes place, the separation line is displaced downstream. This all decreases to a greater or lesser degree the hydrodynamic drag of the sphere.

In the first part of the experiment, it was found that a high turbulence ($\text{Re}_1 \sim 10^5$) of the gas flow alone is the necessary but not sufficient condition for the early crisis of drag to show up. In this case, the time-averaged value of drag coefficient C_d for a sphere about which a free turbulent jet flows is close to C_d for a sphere in a laminar flow (Fig. 5.5) in a wide interval of the transition range $120 < \text{Re} < 530$.

The sufficient condition for the early crisis of drag at a given value of critical Reynolds number Re_c seems to be the fulfillment of criterion (5.1). However, for the interval $10 < \text{Re} < 400$, this supposition calls for quantitative verification.

It seems that the new name of the discovered phenomenon, "The early drag crisis," has the same right to be used as well as the simpler (but less concrete, less certain) name "The drag anomaly," because this phenomenon is very similar to the crisis of drag known at high Reynolds numbers ($\text{Re} > 10^5$): in both cases, the flow about the sphere becomes more streamlined, and the drag decreases markedly.

The interval of the early crisis ($50 < \text{Re} < 200$) and its quantitative aspect (a four- to sevenfold decrease in C_d) are set approximately. As the Reynolds number grew in the range $200 < \text{Re} < 400$, so did drag coefficient C_d of the sphere, approaching standard values in the Rayleigh curve.

Taking account of the early drag crisis for drops is of special importance in simulating two-phase flows in which the velocities of the gas and liquid drops differ considerably, interfacial interaction plays a significant role, and high turbulence may enhance this interaction even more (e.g., such conditions may well take place in sprayers). Otherwise, the results of simulation may turn out to be unreliable, as, for example, those of Ghosh and Hunt [14], with all respect to these authors.

The results presented in this section were previously published partially in papers [10–13, 15, 16].

5.2   Influences of the Geometry and Turbulence of the Gas Flow on the Hydrodynamic Drag of a Streamlined Body

To investigate the effect of coming stream geometry on the hydrodynamic drag of a streamlined body, a numerical experiment was performed in which the gas flow around the sphere was modeled as a free one, and also as flow restricted by cylindrical pipes of different radius, including confuser and diffuser. The results of the calculations led to the conclusion that the restriction of the flow by the walls of the pipe, its narrowing or expansion, can change the magnitude and coefficient of hydrodynamic drag of the body insignificantly: by no more than 30%. This cannot explain the emergence of an early crisis of a body drag, when it decreases by 4–7 times at the same values of Reynolds numbers of the order of 100. A theoretical explanation of this phenomenon is suggested by the influence of strong turbulence of the flow coming to the body.

5.2.1   Summary of Facts About Early Drag Crisis of a Sphere

It has been found experimentally [10, 11] that in a strongly turbulent two-phase flow produced by a spray nozzle, the value of C_d for water drops at Re ~ 100 can be four to seven times smaller (Fig. 5.7) than the conventional values obtained using formula (1.5). At liquid pressure in the nozzle of $P = 5 \cdot 10^5$ Pa, the experimental results are

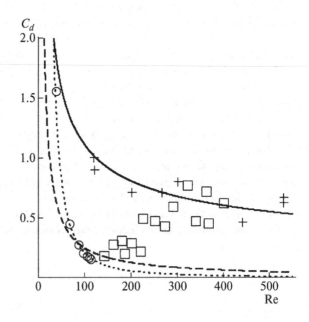

Fig. 5.7 Dependences of drag coefficient C_d for a sphere on Reynolds number Re: the dashed curve is calculated by Stokes formula (1.4); the solid curve corresponds to Klyachko formula (1.5); ○, results of experiment [10] for water drops in a spray at pressure in the nozzle $P = 5 \cdot 10^5$ Pa and the dotted curve approximates these data by formula (5.2); +, results of experiment [13] with a small sphere in a free jet; □, results of the same experiment with a sphere in a flow passing through the confuser [15]

successfully approximated by the dependence (5.2) depicted by the dotted curve in Fig. 5.7. Note that for convenience, Fig. 5.7 is sum of data from Figs. 5.5 and 5.6.

This new effect of the substantial decrease in the value of C_d for transitional Reynolds numbers was called as anomaly or as early crisis of drag [10, 11].

To explain the early crisis of drag [10, 11], the hypothesis 5 was proposed about continuation of the Stokes regime of flow around a drop into the range of transitional Reynolds numbers Re ~ 1–100 due to flow initial strong turbulence (see Sect. 3.4). The basis of this hypothesis was the obvious closeness of experimental values of C_d for drops to the curve describing the Stokes dependence (1.4) continued in the range of Re $\approx$ 1–100.

It was noted earlier in [10, 11] that using the ideas borrowed from theory of the "near-wall" turbulence and connected the existence of a thin viscous sublayer at the drop surface as a part of the turbulent boundary layer [3, 8], a model substantiating this hypothesis could be constructed. This model will be described for the first time in Sect. 5.2.3.

Apart from the abovementioned hypothesis 5, the hypothesis 6 was also formulated concerning the drag crisis emerging due to high initial turbulence of the flow even for Re > 50 analogously to the known phenomenon for a laminar flow around a hard sphere at Re ~ 10^5 (see Sect. 3.4).

The fact that the classical crisis of drag can sometimes emerge "earlier" (i.e., for critical Reynolds numbers Re_c much smaller than 10^5) was considered in detail in [7]. Data about a shift of the drag crisis observed for a sphere as a rule at Re_c ~ 10^5 to the values of Re_c from 2200 to 400 upon an increase in the turbulence degree from 14% to 40%, respectively, was also reported in [7].

In the standard experimental dependence C_d (Re) (sometimes referred to as the Rayleigh curve [1]), the emergence of drag crisis is manifested in a sharp (by several times) decrease in the value of C_d for Re > Re_c and, additionally, as a gradual return of C_d to values close to the standard upon an increase in Re to values several times higher than Re_c, which was observed in some experiments [7]. It can therefore be concluded that in different parts of the wide interval 50 < Re < $3{\cdot}10^5$, the observed crisis is of the local nature.

In work [7], for the interval 400 < Re_c < 2200 on base of experimental data, the dependence (5.1) was derived, which connects the decrease in the critical Reynolds number Re_c with increasing in turbulence relative intensity ε_1 of the flow around a sphere. The authors of work [7] believed that the crisis of drag in the interval 1 < Re < 100 is hardly possible because in accordance with formula (5.1) turbulence relative intensity ε_1 in this case must be higher than 40%, which is hardly probable.

It can be seen from Fig. 5.7, however, that anomaly (or early crisis) of the drag for drops in a spray flow of a nozzle has been also discovered relatively recently in the range of Reynolds numbers 1 < Re < 100 [10, 11]. According to estimates made in these works, the value of ε_1 was 30% and higher.

In an attempt at experimental verification of the early drag crisis for a single rigid sphere [13], it was found that this phenomenon is not observed in a free gas jet flowing around the sphere but was observed when the sphere was at the outlet of a confuser (see Fig. 5.7). This led to the assumption that along with strong turbulence, the flow geometry probably also can affect the hydrodynamic drag.

To verify this hypothesis, we planned and performed a numerical experiment in which the gas streams were simulated flowing around a sphere and having a different geometry, namely, homogeneous flow, as well as flows in circular tubes of different radius, in a confuser, and in a diffuser.

5.2.2   Description and Results of Numerical Experiment

The gas flow was described analytically by the continuity equation

$$\partial \rho / \partial t + \operatorname{div}(\rho \mathbf{V}) = 0 \qquad (5.4)$$

and the Navier–Stokes equation

$$\partial \mathbf{V} / \partial t + (\mathbf{V} \nabla)\mathbf{V} = -\nabla P / \rho + \nu \Delta \mathbf{V}. \qquad (5.5)$$

Here, $\nu = \mu / \rho$ is the kinematic viscosity of the gas. The gas pressure P is related to gas density ρ by the Poisson adiabatic equation $dP = \gamma P / \rho d\rho$, where $\gamma = 1.40$ is the adiabatic exponent for air.

To choose the coordinates system and the shape of the body in the flow, which are optimal for simulation, we took into account the following considerations. If the spherical system of coordinates is used for solving the problem of the gas flow around a sphere in a circular cylindrical or conical tube, it is convenient to specify the boundary of the gas flow on the surface of the sphere, but it is extremely inconvenient to do that on the surface of the tube because the generatrix line of the tube surface does not pass through the points of the computational mesh. The inverse situation takes place if we use the cylindrical system of coordinates. For this reason, a compromise variant was chosen for simplicity. We used a quasi-sphere (i.e., a body of revolution with an octagonal axial section, whose sides can be obtained by connecting the points of the computational mesh) as the body in the flow in the cylindrical system of coordinates (Fig. 5.8). This body could be fully inscribed into a sphere of radius $R = h \, (45)^{1/2} \sim 6.71h$ (h is the mesh size) located on the axis of the gas flow in the middle of the computational domain (see Fig. 5.8).

Equations (5.4) and (5.5) in the chosen system of coordinates were written in the finite-difference form using the explicit Lax–Wendroff algorithm [17]. In calculating the free flow around the body or the flow in a circular cylindrical tube, the computational mesh was a rectangular domain with 46 points along the flow axis (along the z coordinate) and 23, or 29, or 35 points along radius r. In calculating the flow in a conical tube, its generatrix line was inscribed into a suitable rectangular domain so that it passed through some points of the computational mesh (Fig. 5.8).

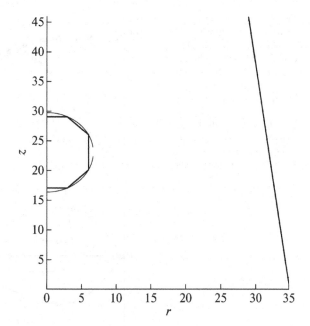

Fig. 5.8 Schematic diagram of the computational domain with the boundaries of solids: a quasi-sphere and a confuser with a radius decreasing from 35 to 29 grid steps of size h [15]

The difference analogs of Eqs. (5.4) and (5.5) were supplemented with the appropriate boundary conditions and were solved numerically using the relaxation method for finding the steady-state solution. In particular, the boundary condition at the solid walls was $\mathbf{V} = 0$ (zero velocity of the gas).

The hydrodynamic drag force acting on the body in direction of the flow was determined from the computed fields of gas velocities and pressures by integrating the stresses over surface of the body

$$F = \int \left(P n_z - 2\mu \partial V_z / \partial z n_z - \mu (\partial V_z / \partial r + \partial V_r / \partial z) n_r \right) df \qquad (5.6)$$

where V_z and V_r are the projections of the gas velocity onto the axes of the cylindrical system of coordinates, $n_z = -\cos\theta$ and $n_r = -\sin\theta$ are the projections of the inwards normal vector to the surface of the body in the flow, and θ is the polar angle between radius vector $\mathbf{r}$ of the given surface point of the body and the z axis. Then, drag coefficient C_d expressed from formula (1.3) was calculated. The calculated gas velocity profiles are shown in Figs. 5.9 and 5.10 in the tube axial section.

It can be seen from Fig. 5.9 that the velocity profiles obtained for the Stokes flow regime (Re = 0.08) are almost symmetric; the function of axial velocity V_z $(z - z_0) = V_z (z_0 - z)$ is even, while the function of radial velocity $V_r (z - z_0) = -V_r$ $(z_0 - z)$ is odd relative to cross section $z_0 = 23h$, in which the center of the quasi-sphere is located. It is well known that the same is observed for the free flow around a sphere in the Stokes regime [3, 8]. This can be regarded as a confirmation of the correctness of our calculations. When the quasi-sphere is in the flow at a transitional

Fig. 5.9 Radial velocity profiles of a gas flowing around a body in a circular cylindrical tube for a Stokes regime of flow at Re = 0.08: o − j = 0 – at the input boundary of the mesh; □, j = 12; ◊, j = 23, at a plane with the center of the quasi-sphere; x, j = 34; +, j = 46, at the output boundary of the mesh; in conventional units $V_z \geq V_r$ for all r [15]

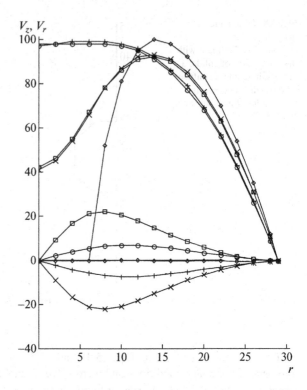

regime (Re = 71.4), such a symmetry is not observed (see Fig. 5.10). On the contrary, in this case, a reverse vortex flow obviously emerges near the rear part (stern) of the body as in the case of the flow around a correct sphere [1].

The results of calculation of drag coefficient C_d are represented in Figs. 5.11, 5.12, 5.13, and 5.14. It is seen from Figs. 5.11 and 5.12 the values of C_d calculated for a body in a free flow at Re < 1 are in conformity with Stokes formula (1.4), while for Re > 5, these values are in agreement with Klyachko formula (1.5). A small (up to 11%) discrepancy between the calculated values of C_d and those given by formula (1.4) or (1.5) can be explained by the deviation of the shape of the body modeling the sphere and by the finite number of the points of the computational mesh.

For a body placed into the flow in a circular cylindrical tube, the calculated values of C_d are slightly lower than those for a free flow. For Re < 1, these values decrease with increasing tube radius (see Fig. 5.11), while for Re > 5, the values of C_d increase (Fig. 5.12). The latter circumstances can be explained by the opposite actions of the following factors: when the tube radius increases (at a constant gas velocity), the pressure drop along the tube axis decreases, and its contribution to the force acting on the body in the flow also decreases; at the same time, the gas flow interacting with the body via viscous friction increases.

For Re < 1, the former factor is more significant, while for Re > 5, the second factor plays the major role. It should be noted that in the Stokes regime (Re < 1), the

Fig. 5.10 Velocity profiles of a gas flowing around a body in a circular cylindrical tube at the transitional regime for Re = 71.4 (notations is the same as in Fig. 5.9) [15]

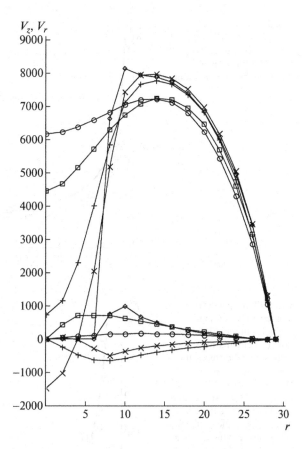

Fig. 5.11 Dependence of drag coefficient C_d on Reynolds number Re in a free flow of a gas and in its flow in a circular tube; dashed curve describes Stokes function (1.4); o, calculation for a sphere in a free gas flow; □, the same for a sphere in a circular tube of radius 23h; +, the same for a tube of radius 29h; ◊, the same for a tube of radius 35h [15]

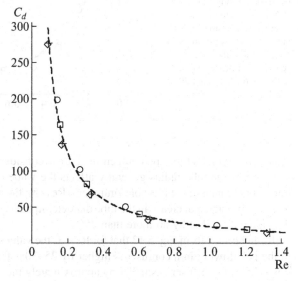

Fig. 5.12 Same as in
Fig. 5.11 for other values of
Re; solid curve describes
Klyachko function (1.5)
(other notations is the same
as in Fig. 5.11) [15]

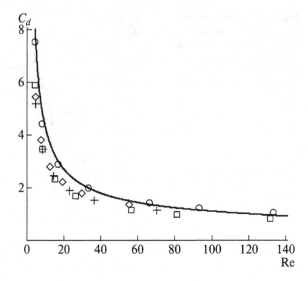

Fig. 5.13 Dependence of
drag coefficient C_d on
Reynolds number Re for a
flow in a circular tube,
confuser, and diffuser; the
dashed curve corresponds to
Stokes function (1.4); +,
calculation for a sphere in a
gas flow in a circular tube of
radius 29h; x, the same for a
sphere in a confuser with a
radius decreasing from 35 to
29h; □, the same for a
sphere in a confuser with a
radius decreasing from 29 to
23h; ◇ ,the same for a
sphere in a diffuser with a
radius increasing from 29 to
35h [15]

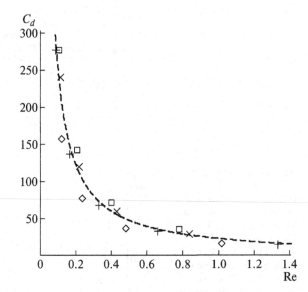

effects of both factors are compensated to a considerable extent; therefore, the variation of the tube radius generally affects the value of C_d insignificantly. In the transitional range of the Reynolds numbers (Re > 5), the effect of the second factor is more significant: a decrease in the tube diameter in the cases considered here reduced the value of C_d, but by no more than 25%.

It can be seen from Fig. 5.13 that for Re < 1, the values of C_d calculated for a body in the gas flowing in a confuser is higher by 25–30% than in a circular tube, while this value for a diffuser is smaller by approximately the same value. For Re > 5, the

Fig. 5.14 Same as in
Fig. 5.7 for other values of
Re; solid curve describes
Klyachko function (1.5)
(other notations are the same
as in Fig. 5.13) [15]

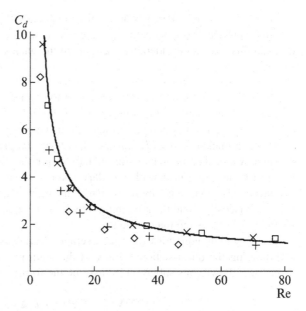

results of calculation of C_d for a body in the confuser are 30% higher than in a circular cylindrical tube and are close to the values given by Klyachko formula (1.5), while for the body in the diffuser, these values are approximately 5–15% smaller than for the circular tube (see Fig. 5.14).

Thus, our numerical experiment shows that the geometry of the flow around a body (e.g., a sphere) can affect hydrodynamic force F and drag coefficient C_d of the body, increasing or decreasing the values of these quantities (however, by not more than 25–30%). This circumstance itself cannot explain the emergence of the early crisis of drag, in which these values are reduced by a factor of 4–7 [10, 11]. In the experiment with a solid sphere, this can be explained by a considerable increase in the degree of turbulence ε_1 of the gas flowing through a confuser [13].

5.2.3   Theoretical Explanation of the Early Drag Crisis

Let us suppose that for Re ~ 10–10^2, a stationary sphere is in a strongly turbulent flow, e.g., circular jet characterized by Reynolds number $Re_1 = <V>D\rho/\mu \sim 10^5$ and the kinematic coefficient of turbulent viscosity ν_τ, which can be assumed to be constant within the jet [8] and to estimate by the formula

$$\nu_\tau = \sigma(J/\rho)^{1/2} = \sigma(\pi/4)^{1/2}\nu\,Re_1 = \text{const} \approx 0.02\nu\,Re_1 \sim 2 \cdot 10^3\nu \gg \nu, \quad (5.7)$$

where $J = \pi/4D^2\rho<V>^2 = \text{const}$ is the momentum flux of the jet, D is the characteristic diameter of the jet cross section, $<V>$ is the gas velocity averaged over the cross section, and $\sigma \approx 0.021$ is an empirical constant [8]. It should be emphasized that in accordance with estimate (5.7), the kinematic coefficient of turbulent viscosity ν_τ in this case is considerably higher than the analogous coefficient of physical viscosity ν.

In accordance with the theory of "near-wall" turbulence [3, 8], a turbulent boundary layer is formed at the surface of the sphere, in which the momentum transfer is determined by the joint action of the turbulent and physical viscosities $\nu_\Sigma = \nu_\tau(y) + \nu$. The first term ν_τ varies with distance y from the sphere surface and over a small thickness ($\delta \ll d$) of a boundary layer sharply increases from zero to the maximal value determined by formula (5.7) for the flow core. To simplify the physical pattern, the near-wall boundary layer is often assumed to be formed by two layers, viz., a viscous sublayer in which the momentum transfer is determined by physical viscosity ν and the remaining flow in which quantity $\nu_\tau = $ const defined by formula (5.7) plays the major role [8].

The Reynolds equations for time-averaged variables velocity $\mathbf{V}$ and pressure P describing the quasi-stationary flow in the main part of the turbulent boundary layer (over the viscous sublayer) [8] assume the form

$$(\mathbf{V}\nabla)\mathbf{V} = -\nabla P/\rho + \nu_\tau \Delta \mathbf{V}, \tag{5.8}$$

$$\mathrm{div}\mathbf{V} = 0. \tag{5.9}$$

Equation (5.8) obviously differs from the stationary Navier–Stokes equation for an incompressible fluid only by the substitution of ν for ν_τ. The ratio of the first (convective) term to the third (viscous) term in Eq. (5.8) is determined (with allowance for estimate (5.7)) by small quantity $\mathrm{Re}_\tau = Vd\rho/\nu_\tau = \mathrm{Re}\nu/\nu_\tau < 1$, which differs from the Reynolds number Re by the substitution of ν for ν_τ.

For this reason, we can disregard the first term in Eq. (5.8) in view of its smallness (as in the classical Stokes problem of the flow of a viscous liquid around a sphere for $\mathrm{Re} \ll 1$), and this equation can be written in the form

$$\nu_\tau \Delta \mathbf{V} - \nabla P/\rho = 0. \tag{5.10}$$

In the abovementioned Stokes problem, the **rot** operator is sometimes applied to the analog of the Eq. (5.10), which contains ν instead of ν_τ. This gives the equation

$$\Delta\,\mathbf{rot}\mathbf{V} = 0, \tag{5.11}$$

which does not contain variable of pressure P and which can be solved (as in [3]) together with continuity Eq. (5.9).

Since the resultant system of Reynolds Eqs. (5.9 and 5.11) for the main part of the turbulent boundary layer around the sphere does not differ (to within substitution of ν_τ for ν) from the analogous system of equations in the classical Stokes problem, the procedure of solving Eqs. (5.8 and 5.11) and the result for $\mathbf{V}$ are the same as in the

Stokes problem. In the spherical system of coordinates (r, φ, θ) with the origin at the center of the sphere, this solution is axially symmetric and has the form

$$V_r = V_\infty \cdot \cos \theta \cdot \left[1 - 3R/(2r) + R^3/(2r^3)\right], \qquad (5.12)$$

$$V_\theta = -V_\infty \cdot \sin \theta \cdot \left[1 - 3R/(4r) - R^3/(4r^3)\right], \qquad (5.13)$$

where $R = d/2$ is the radius of the sphere [3, 8].

It should be noted that solution (5.12 and 5.13) also holds for a thin region of the viscous sublayer adjoining the surface of the sphere, in which the flow is described by the same Eqs. (5.9 and 5.11) (to within the inverse substitution of ν for ν_τ). In accordance with the near-wall turbulence theory [8], the velocity components in this region (for $r - R = y \to 0$) are connected by the relation

$$V_\theta \sim y/R \gg V_r \sim (y/R)^2 \to 0,$$

and the momentum transfer is determined by conventional kinematic viscosity ν.

In the classical Stokes problem of the viscous flow around a sphere for Re $\ll 1$, the pressure can be determined from solution (5.12 and 5.13) using the analog of Eq. (5.10) containing ν instead of ν_τ:

$$P = P_\infty - 3/2\nu\rho V_\infty R \cos\theta/r^2. \qquad (5.14)$$

And integrating the stresses over the surface of the sphere, we can determine the hydrodynamic drag force acting on the sphere:

$$F = \int \left(-P\cos\theta + 3/2\nu\rho V_\infty \sin^2\theta/R\right)df = 6\pi R\mu V_\infty \qquad (5.15)$$

Using formula (1.3), we can now find the drag coefficient C_d, which has the form (1.4) in the Stokes problem. The same procedure with the same results (5.15) and (1.4) can be carried out in the problem of the strongly turbulent gas flow around the quasi-sphere considered here.

It should be noted that the shear stress on the surface of the sphere, which is taken into account by the second term in the integrand of formula (5.15), in the viscous sublayer of the turbulent boundary layer, is determined correctly by coefficient of conventional kinematic viscosity ν but not of turbulent viscosity ν_τ. The same coefficient ν is presented in formula (5.14) as well as in the expression for Reynolds number Re $= V_\infty \rho d/\nu$ in formula (1.4).

Conversely, we must substitute ν_τ for ν in formula (5.14) for the external part (above the viscous sublayer) of the turbulent boundary layer around the sphere. This means that upon a transition from the viscous sublayer to the outer part of the

Fig. 5.15 Dependence of drag coefficient C_d on Reynolds number Re: dashed curve corresponds to Stokes formula (1.4); solid curve is calculated by Klyachko formula (1.5); o, experiment [10] with water drops in a spray for pressure $P = 5 \cdot 10^5$ Pa; x, the same for $P = 3 \cdot 10^5$ Pa; $\Diamond$, the same for $P = 9 \cdot 10^5$ Pa; $\Box$, experiment [13] with a sphere in a flow passing through the confuser [15]

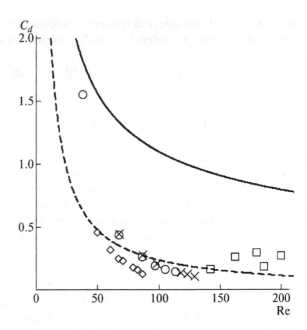

boundary layer, pressure P may abruptly change (or continuously in the intermediate transition layer).

In Fig. 5.15, Stokes dependence (1.4) is continued onto the range of Re > 1 and compared with the experimental results for the early crisis of drag for drops and of a single solid sphere in strongly turbulent flows. In supplement to the results given in Fig. 5.7, the data on drag coefficient C_d for drops in Fig. 5.15 are given not only for liquid pressure in the nozzle $P = 5 \cdot 10^5$ Pa but also for pressures of $3 \cdot 10^5$ and $9 \cdot 10^5$ Pa. It can be seen that theoretical dependence (1.4) is in good agreement with the data on the drag crisis of spherical particles in the transitional region of Re $= 50$–200.

5.2.4   Comparison of Drag Crises of Spheres for Different Values of Re_c

Thus, we have established that for a high degree of gas flow turbulence ($\varepsilon_1 > 30\%$) and for a considerable turbulent viscosity $\nu_\tau \gg \nu$, the time-averaged flow around a body for Re ~ 50–200 is similar to the Stokes flow which is not separated from the body surface. In this case, according to Fig. 5.15, drag coefficient C_d of the body, which is approximately equal to the values defined by Stokes formula (1.4), is much smaller than the values calculated by Klyachko formula (1.5) for a laminar flow. It is this difference that was detected in the effect of early drag crisis [1, 2, 7].

In its physical meaning (the decrease in the hydrodynamic drag of the body by several times due to the improvement of the flow conditions), this new phenomenon is analogous to the familiar effect of drag crisis for a sphere in a laminar flow for $Re \sim 10^5$. In the familiar classical crisis, the separation of the turbulent boundary layer from the surface of the sphere takes place with downstream displacement of the separation line [3, 8]. However, in the early crisis, there is no separation of the turbulent boundary layer, at least, for C_d close to the minimum value (see Fig. 5.15).

The initial turbulence of the incoming flow to a body displaces (due to natural or artificial preliminary turbulization) the drag crisis toward smaller values of Re_c. When the turbulence intensity decreases from 0.5% to 2.5%, the value of Re_c decreases from $2.7 \cdot 10^5$ to $1.25 \cdot 10^5$ [8].

As noted above, the cases of emergence of the drag crisis in the intermediate interval $400 < Re_c < 2200$ between $Re_c \sim 100$ and $Re_c \sim 10^5$, corresponding to the early and classical crises, were described in [7]. The degree of flow turbulence (40% $> \varepsilon_1 > 14\%$, respectively) is also intermediate: it is higher than 5%, which still corresponds to the classical crisis [7], but smaller than the value at which the early crisis takes place (this value is not determined exactly). In these intermediate cases, as well as in the classical crisis, the turbulent boundary layer is apparently separated, and the line of separation is displacement downstream.

As the basic characteristics of any of the above drag crises, we can take critical Reynolds number Re_c, at which the value of $C_d = 0.3$ is attained, and the minimal value of the drag coefficient $\min(C_d) \sim 0.1$ (according to the results of [7, 10]).

The value of Re_c is determined by the degree of turbulence of the flow oncoming onto the sphere; the quantitative measure of this quantity is associated with the following three characteristics: Reynolds number Re_1 of the flow, turbulence intensity s, and kinematic turbulent viscosity ν_τ. The relation between some of these characteristics (e.g., ν_τ and Re_1) for a free turbulent jet is known (see relation (5.7)). In some cases of classical and "intermediate" crises, relation (5.1) between Re_c and ε_1 has been established. The validity of this relation for the early crisis is doubtful because, as can be seen from Fig. 5.15, $C_d = 0.3$ for water drops in the spray for $Re_c = 60$–80; accordingly, the values of $\varepsilon_1 = 87$–75% obtained using formula (5.1) are incredibly high. In addition, it is known that the value of ε_1 varies along the radius of the tube or jet; it is minimal at the flow axis and strongly increases (by two to three times) at the flow boundary (at the tube wall and at the boundary of the free jet) [1, 8]. In spite of this, the early drag crisis was observed on the axis of a two-phase jet as well as near its boundary for approximately identical values of Re_c. In contrast to ε_1, the value of ν_τ can be treated as constant along the jet radius in accordance with relation (5.7) [8].

A common feature of the classical and early drag crises (and probably of intermediate crises) is the high value of number $Re_1 \sim 10^5$, which characterizes the flow oncoming onto the body. Unless special measures have been taken for stabilization and laminarization of a flow (e.g., by using a tube with a smoothly narrowing inlet), then this circumstance alone can ensure a high degree of the flow turbulence in the tube as well as in the free jet.

The high degree of turbulence in the spray of a nozzle ($\varepsilon_1 > 30\%$) proved to be sufficient for the emergence of the early drag crisis for the drops [10, 11]. This is apparently a distinguishing feature of such two-phase flows. For a single solid sphere in a turbulent jet from a fan, the degree of turbulence for $\mathrm{Re}_1 \sim 0.5 \cdot 10^5$ was insufficient for the emergence of the early crisis. However, such a crisis was observed when a small sphere was at the outlet of a confuser placed in the path of the jet, where the turbulence intensity was apparently higher than in the free jet [13].

In observation the crises of drag in the intermediate range of $2200 > \mathrm{Re}_c > 400$, the gas flow in a tube was additionally turbulized by using special grids, and the value of ε_1 was artificially increased to 14–40%, respectively, by organizing the motion of particles in the turbulent flow with the same direction and a reduced relative velocity [7]. In this case, the natural relation connecting the turbulence characteristics Re_1, ν_τ, and ε_1 is apparently lost. Therefore, the validity of relation (5.1) between Re_c and ε_1 for various values of Re_c remains dubious. The relations of turbulent characteristics with one another (Re_1 and ε_1; ν_τ and ε_1) and with a crisis characteristic (ν_τ and Re_c; Re_1 and Re_c) also remain unclear.

5.2.5   Conclusions About Influences of Geometry and Turbulence of Gas Flow

Thus, the numerical experiment described in Sect. 5.2.2 has demonstrated that the geometry of the flow around a body (e.g., a sphere) may affect hydrodynamic force F and drag coefficient C_d of the body, increasing or decreasing the values of these quantities by less than 25–30%. This circumstance cannot explain the emergence the early crisis of drag, in which these values decrease by four to seven times [10, 11]. In experiments with a solid sphere, this decrease can be due to an increase in the degree of turbulence of the gas flowing through a confuser [13].

The theoretical explanation the early crisis of drag for a sphere in a flow with a high turbulent viscosity given in Sect. 5.2.3 is based on analogy with the Stokes regime of the flow. The values of C_d obtained using the proposed theoretical concepts are in satisfactory agreement with experiment (see Fig. 5.15).

In Sect. 5.2.4, comparative analysis is presented for the known crises of drag for a sphere for various values of critical Reynolds number Re_c. It is noted that interesting and important relations between turbulent and crisis characteristics (Re_1 and ε_1; ν_τ and ε_1; ν_τ and Re_c; Re_1 and Re_c) have not been established as of yet.

Some of the results presented in this section were previously published in articles [13, 15].

5.3 Calculations of the Laminar and Strongly Turbulent Flow About a Sphere and Its Hydrodynamic Drag Force

To investigate the influence of a strongly turbulent flow on the hydrodynamic drag of a body and occurrence of the early crisis of drag, a numerical experiment is conducted in which a free gaseous flow about a sphere is simulated for two cases, namely, as a laminar flow and as a strongly turbulent flow. Turbulence is simulated by using a high value of kinematic coefficient of turbulent viscosity. Calculation data lead us to conclude that the early crisis of drag for Reynolds numbers near 100, which shows up as a considerable (four- to sevenfold) decrease in the hydrodynamic force and drag coefficient of the body, can be explained by the strong turbulence of the oncoming flow.

5.3.1 Introduction to the Task

In [10, 11], it was found experimentally (Fig. 5.15) that in a strongly turbulent flow with Re ~ 100, the value of C_d for drops may be four to seven times smaller than that given by formula (3.4). At the same time, the early crisis of drag was not observed in a free gas flow about a solid sphere but was observed in a jet flowing through a confuser [13].

Note that data points presented in Fig. 5.15 are well approximated by a dashed curve corresponding to Stokes formula (3.3).

Numerical experiment [15] did not validate the supposition that the geometry of an incoming flow may have a considerable influence on the hydrodynamic drag of a body in the flow. It was also hypothesized that the early crisis of drag of a spherical particle arises because of the influence of the gas flow strong turbulence. The flow turbulence in a confuser may be stronger than that for a free jet and become sufficient for the early crisis to occur [13, 15].

To verify this hypothesis, a numerical experiment was conducted in which free laminar gaseous flow about a sphere was simulated as well as strongly turbulent flow.

5.3.2 Simulation of a Laminar Flow About a Sphere

Usually (see Sect. 5.2) the mathematical model of the gas flow includes the continuity Eq. (5.4) and the Navier–Stokes Eq. (5.5).

The gas pressure and density are related by the Poisson's adiabatic equation, $dP = \gamma P d\rho/\rho$, where γ is the adiabatic exponent. In our calculations, we used the value of $\gamma = 1.4$ for air.

Equations (5.4) and (5.5) were initially written in the spherical coordinate system and then were presented in the form of finite differences using the Lax–Wendroff explicit scheme [17]. The computational domain in the form of a semiring had 50 points along radius $r_j = jh$ (where $j = 10$–59 is the layer no.) and 33 points ($i = 0$–32) over polar angle $\theta_i = 0 - \pi$ between the z polar axis and radius vector $\mathbf{r}$ of a given point. The center of symmetry of the domain coincided with the center of a sphere with radius $R = 10h$ where h is the mesh spacing along r. The equations of hydrodynamics complemented by appropriate boundary and initial conditions were solved by the relaxation method of finding a steady-state solution.

From the calculated velocity and pressure fields of a gas, we found force acting on the sphere by integrating stresses over its surface:

$$F = \int \left(-P\cos\theta + 3/2\nu\rho V_\infty \sin^2\theta/R \right) df. \tag{5.16}$$

Then, we calculate drag coefficient C_d of the sphere by using similar (1.3) well-known formula:

$$F = C_d S \rho V^2 / 2. \tag{5.17}$$

Here and after, we use the following notation: $V = V_\infty$ is the gas velocity far from the sphere, $S = \pi d^2/4$ is area of the midsection of a sphere, d is its diameter, ρ is the gas density, and μ and $\nu = \mu/\rho$ are the dynamic and kinematic viscosities of the gas, respectively.

Calculation data are presented in Figs. 5.16, 5.17, 5.18, 5.19, 5.20, and 5.21.

Figures 5.16 and 5.17 shows the profiles of the radial V_r and tangential V_θ components of the normalized gas velocity $\mathbf{V}(r, \theta)/V_\infty$. The values of angular coordinate $\theta_i = \pi i/32$ are plotted on the abscissa. Symbols in Figs. 5.16 and 5.17 are data obtained by numerical simulation of the laminar gas flow about the sphere in the Stokes regime (Re = 0.25 < 1) for spherical layers with numbers $j = 16, 22$, and 44 along the radius. For the same layers, lines depict profiles obtained from the known analytical solution to the Stokes problem [8] normalized in the same way:

$$V_r = \cos\theta \cdot \left[1 - 3R/(2r)/ + R^3/\left(2r^3\right) \right], \tag{5.18}$$

$$V_\theta = -\sin\theta \cdot \left[1 - 3R/(4r)/ - R^3/\left(4r^3\right) \right]. \tag{5.19}$$

In Figs. 5.16 and 5.17, good agreement between both results is seen.

The profiles of velocities V_r and V_θ similar to those presented in Figs. 5.16 and 5.17 are shown in Figs. 5.18 and 5.19, respectively. They are calculated for the case of the laminar gaseous flow about the sphere in the transient regime (Re = 128). In Figs. 5.16, 5.17, 5.18, and 5.19, data for spherical layers with the same numbers j along radius r are marked by the same symbols.

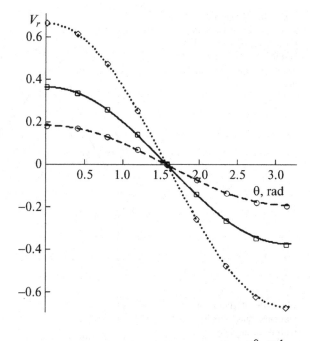

Fig. 5.16 Variation of radial gas velocity V_r with polar angle θ, at Re = 0.25. Number j of the spherical layer along the radius is (O) 16, (□) 22, and (◊) 44 [16]

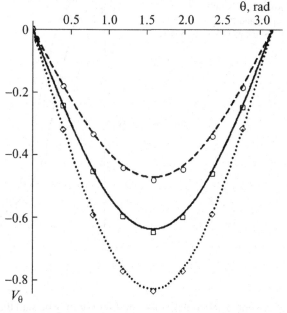

Fig. 5.17 Variation of tangential gas velocity V_θ with polar angle θ_i at Re = 0.25. Numbers j of the layers and the symbols are the same as in Fig. 5.16 [16]

It should be particularly noted that the curves and symbols in Figs. 5.18 and 5.19 correspond to identical profiles obtained by numerical simulation, whereas in Figs. 5.16 and 5.17, the symbols stand for the results of calculations by using the two-dimensional numerical model, and the curves present the analytical solutions

Fig. 5.18 Variation of radial gas velocity V_r with polar angle θ_i at Re = 128. Numbers j of the layers and the symbols are the same as in Fig. 5.16 [16]

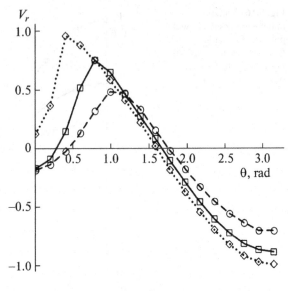

Fig. 5.19 Variation of tangential gas velocity V_θ with polar angle θ_i at Re = 128. Numbers j of the layers and the symbols are the same as in Fig. 5.16 [16]

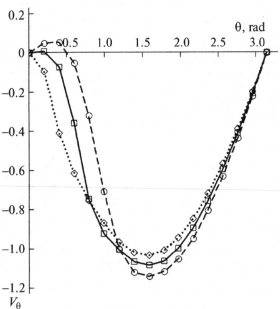

(5.18 and 5.19) to the Stokes problem. It may be said that the curves in Figs. 5.18 and 5.19 are a graphical interpolation of the calculated profiles shown by the symbols.

The results presented in Figs. 5.16 and 5.17 and, respectively, in Figs. 5.18 and 5.19 are seen to differ significantly. In particular, in Figs. 5.18 and 5.19, one can notice a reverse gas flow near the stern part of the sphere: $V_r < 0$ and $V_\theta > 0$ at $j = 16$ and $i < 5$.

Fig. 5.20 Re dependence of the drag coefficient of the sphere in the laminar flow about the sphere in the Stokes regime (Re < 1): (O) results of calculation by the numerical model; solid curve, results of calculation by formula (1.4) [16]

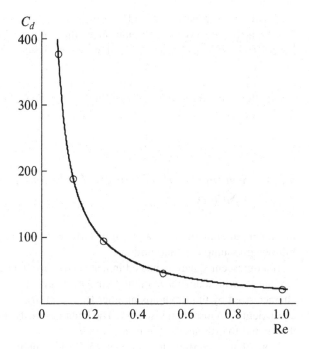

Fig. 5.21 Re dependences of the drag coefficient of the sphere at transient regime of the flow (Re > 1): (o) results for the laminar flow about the sphere calculated by the numerical model; solid line, calculation by formula (1.5); (□) and dashed line, calculation of the strongly turbulent flow about the sphere; and (◊) experimental data [10] for drops in the spray flow at liquid pressure in a nozzle P = 5 atm [16]

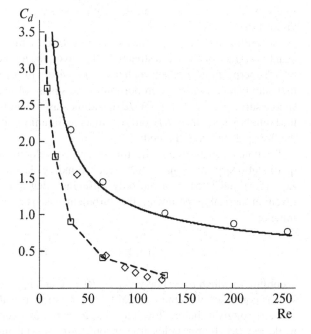

Small circles in Figs. 5.20 and 5.21 refer to dependences of drag coefficient C_d on Re obtained by numerical simulation of the laminar flow about the sphere. Figure 5.20 also shows classical dependence (1.4), which is valid for a Stokes laminar flow about the sphere. Figure 5.21, in turn, demonstrates Klyachko dependence (1.5), which well fits the experimental data in the transitional range $2 < \text{Re} < 700$. It is seen that numerical simulation of the laminar flow about the sphere gives results that are in good agreement with earlier known data.

5.3.3   Simulation of a Strongly Turbulent Flow About a Sphere

In numerical simulation of a strongly turbulent flow about a sphere, we used the following assumptions and notions.

If a quiescent sphere is placed in a strongly turbulent gas flow, for example, in a circular jet with diameter D in the transient range ($\text{Re} \sim 10\text{--}100$), the stream can be characterized by the Reynolds number $\text{Re}_1 = <V>D\rho/\mu \sim 10^5$ and by the kinematic coefficient ν_τ of turbulent viscosity. The latter can be assumed to be invariable within the jet and to estimate by the formula (5.7).

Note that according to estimate (5.7), kinematic coefficient ν_τ of turbulent viscosity in the given case considerably exceeds the ordinary coefficient of physical viscosity ν.

According to the theory of near-wall turbulence, a turbulent boundary layer forms near the surface of a body in stream. In this layer, the momentum transfer is governed by a "superposition" of the turbulent and physical viscosities, $\nu_\Sigma = \nu_\tau(y) + \nu$ [8]. The first summand ν_τ varies in proportion to the square of distance $y = r - R$ from the sphere surface, $\nu_\tau = (0.4y)^2|\partial V_\varphi/\partial y|$, and increases sharply over small thickness δ of the boundary layer ($\delta \ll R$) from zero to a maximum value given by formula (5.7) for the flow part far from the body.

Reynolds equations for the time-averaged velocity and pressure describing a quasi-stationary flow in a turbulent boundary layer have the same form as Eqs. (5.4) and (5.5) with the only difference that ν is substituted by ν_Σ. In the given numerical experiment, the turbulent viscosity was approximated by the function

$$\nu_\tau(y) = \nu_\tau(\infty)(1 - R/r)^2 = 2000\nu(1 - R/r)^2, \tag{5.20}$$

according to which $\nu_\tau \to 0$ at $y = (r - R) \to 0$ and $\nu_\tau \to \text{const} = 2000\nu_\tau$ at $y \to \infty$. This is consistent with the ideas of the theory of near-wall turbulence.

In a strongly turbulent flow, the pressure variation over its cross section is usually neglected [8]. If this holds true in our case as well until the boundary layer is separated, neglect of the first summand in the integrand of formula (5.16) is valid.

Fig. 5.22 Variation of radial gas velocity V_r with polar angle θ, at Re = 128. Shown are calculation data for the strongly turbulent flow about the sphere in the layers with numbers j = (x) 11, (o) 16, ($\square$) 22, and ($\Diamond$) 44. The corresponding curves are the velocity profiles for the spherical layers with the same numbers j obtained from solutions (5.18 and 5.19) of the Stokes problem [16]

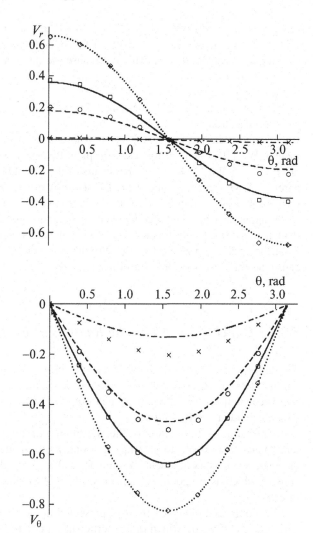

Fig. 5.23 Variation of tangential gas velocity V_θ with polar angle θ_i at Re = 128. Numbers j of the layers, symbols, and curves are the same as in Fig. 5.22 [16]

Symbols in Figs. 5.22 and 5.23 stand for the profiles of velocities V_r and V_θ, respectively. These profiles were obtained by numerically simulating a strongly turbulent flow about a sphere in the transitional range at Re = 128. Curves depicted in Figs. 5.22 and 5.23 are theoretical profiles (5.18 and 5.19) for the same layers (j = 11, 16, 22, 44), which are the solutions to the Stokes problem extended into the transient range of Re numbers (they are shown for comparison).

It follows from Figs. 5.22 and 5.23 that the two distributions of gas velocity component V_r over angle θ, namely, those obtained using the numerical model and theoretical ones (by Stokes theory), differ insignificantly in all spherical layers of radii $r_j = jh$. The two distributions of component V_θ derived in the same way differ only in the boundary layer, i.e., near the sphere (j = 11–16) over distances of

0.1–0.6 of radius R from it (Fig. 5.23). Remarkably, in the turbulent flow about the sphere, the normalized gas velocity in the boundary layer is higher than in the conditions of Stokes regime. This is because the influence of physical viscosity ν, which value is much smaller than turbulent viscosity ν_τ in the incoming flow, becomes more substantial near the sphere.

Calculated data for drag coefficient C_d of the sphere obtained by numerically simulating a strongly turbulent flow about a sphere ($Re_1 \sim 10^5$) are shown by small squares and a dashed line in Fig. 5.21. Rhombuses in this figure stand for experimental data [10] obtained for water drops in the spray flow at liquid pressure in a nozzle $P = 5$ atm. At $Re > 50$, the results of the calculation are seen to be in good agreement with experimental data. This confirms that our conception of the early crisis of hydrodynamic drag in a strongly turbulent flow is correct.

Disagreement between the calculated and experimental data at $Re < 50$ may be explained as follows. In this range of Re numbers corresponding to short distances from the spraying nozzle [10], the turbulence of the gas flow was insufficient to cause the early crisis of drag for drops.

5.3.4 Conclusions About the Reasons of Early Crisis of Drag

The numerical experiment showed that the proposed algorithm for calculating a laminar gaseous flow about a sphere yields values of V_r, V_θ, and C_d that are consistent with known theoretical data obtained at $Re < 1$ and with experimental data obtained at $1 < Re < 400$.

In addition, a combination of the algorithm with the theory of near-wall turbulence made it possible to simulate a strongly turbulent flow about a sphere and calculate velocity components V_r and V_θ and also the drag coefficient C_d of the sphere, which is in good agreement with the experimental data for early crisis of drag.

This confirmed the correctness of presented in the paper [15] explanation of the early crisis of drag by which drag coefficient C_d in a turbulent flow about a sphere decreases four- to sevenfold compared with the case of a laminar flow. This effect is attributed to the strong initial turbulence of a flow about a spherical particle. When turbulent viscosity ν_τ of the flow is high, the flow conditions and the profiles of the time-averaged gas velocities become similar to those observed in the Stokes regime of flow at $Re < 1$. In addition, when $\nu_\tau(\infty)$ is much higher than physical viscosity ν, which plays a key role near the sphere, the drag coefficient of the sphere decreases by several times.

The results presented in this section were previously published in paper [16].

References

1. Torobin, L. B., & Gauvin, W. H. (1959). *Canadian Journal of Chemical Engineering, 37*, 129.
2. Schlichting, H. (1968). *Boundary layer theory* (6th ed.). New York: McGraw-Hill. Nauka, Moscow, 1974.
3. Landau, L. D., & Lifshitz, E. M. (1988). *Course of theoretical physics* (Fluid mechanics) (Vol. 6). Moscow: Nauka. Pergamon, New York, 1987.
4. Belotserkovskii, O. M., & Davydov, Y. M. (1982). *Method of large particles in gas dynamics: Numerical experiments*. Moscow: Nauka.
5. Nigmatulin, R. I. (1987). *Dynamics of multiphase media*. Moscow: Nauka. Hemisphere, New York, 1991.
6. Brounshtein, B. I., & Fishbein, G. A. (1977). *Hydrodynamics of mass and heat transport in dispersed systems*. Leningrad: Khimiya.
7. Torobin, L. B., & Gauvin, W. H. (1960). *Canadian Journal of Chemical Engineering, 38*, 189.
8. Loitsyanskii, L. G. (1978). *Mechanics of liquids and gases*. Moscow: Nauka. Pergamon, Oxford, 1972.
9. Simakov, N. N. (1987). Dissertation, Yaroslavl Polytechnic Institute, Yaroslavl.
10. Simakov, N. N. (2004). Crisis of Hydrodynamic Drag of Drops in the Two-Phase Turbulent Flow of a Spray Produced by a Mechanical Nozzle at Transition Reynolds Numbers. *Technical Physics, 49*, 188.
11. Simakov, N. N., & Simakov, A. N. J. (2005). Anomaly of gas drag force on liquid droplets in a turbulent two-phase flow produced by a mechanical jet sprayer at intermediate Reynolds numbers. *Applied Physics, 97*, 114901.
12. Simakov, N. N. (2002). Numerical simulation of a two-phase flow in the spray stream produced by the nozzle. *Izvestiia Vysshykh Uchebnykh Zavedenii. Khimiya and KhimicheskayaTekhnologiya (News of universities. Chemistry and chemical technology), 45*(7), 125.
13. Simakov, N. N. (2010). Experimental Verification of the Early Crisis of Drag Using a Single Sphere. *Technical Physics, 55*, 913.
14. Ghosh, S., & Hunt, J. C. R. (1994). *Proceedings of the Royal Society of London, Series A, 444*, 105.
15. Simakov, N. N. (2011). Effect of the gas flow geometry and turbulence on the hydrodynamic drag of a body in the flow. *Technical Physics, 56*, 1562.
16. Simakov, N. N. (2013). Calculation of the flow about a sphere and the drag of the sphere under laminar and strongly turbulent conditions. *Technical Physics, 58*, 481.
17. Potter, D. E. (1973). *Computational physics*. New York: Wiley. Mir, Moscow, 1975.

Chapter 6
Calculation of Drag Coefficient of a Sphere and Heat Transfer from It to a Gaseous Flow

The hypothesis about the influence of the early drag crisis of sphere on its heat exchange with gas was confirmed by mathematical modeling. First, the numerical simulation of the gas flow around the sphere in a cylindrical channel was carried out with the calculation of the drag coefficient of sphere and heat transfer from it to a gas. Second, the same was done for the case of flow around the sphere by a free gas stream, both laminar and strongly turbulent. In the latter case, it was found that the early crisis of drag for the sphere is accompanied by a crisis of its heat exchange with gas. In addition, the numerical simulation of the heat exchange of a drop of liquid with a gas stream was carried out without taking into account its evaporation.

6.1 Case of Gaseous Flow Around a Sphere in a Cylindrical Channel

A numerical experiment on the simulation of heat transfer from a sphere to a gaseous flow in a cylindrical channel in the Stokes and transient regime of a flowing has been described. Radial and axial profiles of the gas temperature and the dependences of drag coefficient C_d of the body and Nusselt number Nu on Reynolds number Re have been calculated and analyzed. The problem of the influence of the early crisis of drag on the heat transfer from a sphere to gas flow has been considered. The estimation of this phenomenon has shown that the early crisis of drag for a sphere in a strongly turbulent flow causes a decrease in heat transfer from the sphere to the gas by three to six times (in approximately the same proportion as for its drag coefficient).

6.1.1 Introduction: Early Crisis of Drag for a Sphere

In many technological processes, spraying a liquid in a gas, e.g., using spray nozzles, is used to increase the area of interfacial surface for intensifying the heat and mass transfer between phases.

In this case, drops with average diameter $d \sim 10^{-4}$ m are formed. For these small sizes and the large difference between dynamic viscosities of the droplet liquid and the gas flow around drops (approximately by 60 times for water and air), their deformation and the internal flowing of the liquid in them can be neglected, and they can be considered to be small hard spheres.

Indeed, it is well known (e.g., from the monograph [1]) that for small values of the Weber criterion We $\ll 1$, the shape of the drops deviates insignificantly from the spherical shape. For drops with radius $R_d = d/2$ of water sprayed by a nozzle in air, we have the relations

$$\text{We} = \rho(V_\infty)^2 R_d/\sigma = 1.2 \cdot 15^2 \cdot 0.5 \cdot 10^{-4}/\left(73 \cdot 10^{-3}\right) \approx 0.2 \ll 1 \qquad (6.1)$$

where ρ is the air density, V_∞ is the velocity of drops relatively to the gas, and σ is the surface tension for water. Therefore, the deformation of drops can be neglected.

The following formula is given in the same monograph [1]

$$C_d = (\mu^* C_{d\infty} + C_{d0})/(1 + \mu^*), \qquad (6.2)$$

which expresses hydrodynamic drag coefficient C_d of a drop in terms of the same coefficients $C_{d\infty}$ for a hard sphere and C_{d0} for a gas bubble and the ratio $\mu^* = \mu_L/\mu$ of dynamic viscosities μ_L of the droplet liquid and μ of the gas. For water drops in air, $\mu^* \approx 60$ and, according to formula (6.2), $C_d \approx C_{d0}$ "with an error within 5% for Re < 100" [1]; i.e., the liquid circulation inside a small drop changes the drag coefficient only slightly compared to a hard sphere.

Analyzing the processes with the spraying of a liquid, we must calculate the hydrodynamic drag force of a drop in the gas flow as follows:

$$F = C_d S \rho (V_\infty)^2/2. \qquad (6.3)$$

For this we must know relative velocity V_∞ of the gas flow at a large distance from the drop, its drag coefficient C_d, and the area $S = \pi d^2/4$ of the midsection of the spherical drop.

For a sphere in a laminar gaseous flow with small Reynolds numbers Re $= Vd\rho/\mu \ll 1$, the following well-known Stokes formula can be used

$$C_d = 24/\text{Re} \qquad (6.4)$$

while in the transient range $2 < \text{Re} < 700$, the Klyachko dependence

$$C_{\mathrm{d}} = 24/\mathrm{Re} + 4/\mathrm{Re}^{1/3} \qquad\qquad (6.5)$$

is valid. These dependences approximate the experimental data generalized by the Rayleigh curve in the above ranges quite well [2, 3].

It was shown experimentally [4, 5] (Fig. 5.15) that, in a strongly turbulent flow with Re $\approx$ 100, the value of C_{d} for drops may be four to seven times smaller than the conventional values determined by formula (6.5). The same early drag crisis was observed on a single hard sphere in a gas jet flow through a confuser [6].

It should be noted that experimental points shown in Fig. 5.15 lie near by a dashed curve that corresponds to Stokes formula (6.4).

The numerical experiment [7] did not confirm the assumption that the geometry of an incoming flow may have a considerable influence on the hydrodynamic drag of a body in the flow. As another reason that explains the early crisis of drag for a sphere, a hypothesis was proposed regarding the influence of a strong turbulence of the gas flow, which a confuser can further intensify (compared to a free jet) and make it sufficient for the early crisis which can occur on a single hard sphere [6, 7].

To verify this hypothesis, a numerical experiment was made in which free laminar and strongly turbulent gas flow around a sphere were simulated [8].

In accordance with the well-known analogy that exists in the transfer of momentum, heat, and mass in a nonuniform flow, the profiles of velocities, impurity concentrations, and temperatures in different cross sections of the flow can exhibit a certain similarity [9, 10]. With allowance for this analogy, the drag crisis of a sphere in a strongly turbulent flow must inevitably affect the heat and mass transfer between the sphere and the gas. It is interesting to somehow estimate this influence, e.g., using a numerical simulation of a gas flow around the sphere and taking into account the heat transfer. But firstly, we must construct this model by neglecting the early crisis of drag.

6.1.2 Simulation of Laminar Gaseous Flow Around a Sphere in a Cylindrical Tube Taking into Account Heat Exchange

The mathematical model of the gas flow with low (as compared to velocity V_{s} of sound) velocities V is based on the classical hydrodynamic equations.

Their system includes the continuity equation

$$\partial \rho / \partial t + \mathrm{div}(\rho \mathbf{V}) = 0 \qquad\qquad (6.6)$$

and the Navier–Stokes equation

$$\partial \mathbf{V}/\partial t + (\mathbf{V}\nabla)\mathbf{V} = -\nabla P/\rho + \nu \Delta \mathbf{V} \qquad (6.7)$$

by analogy with the model described in Sect. 5.2.

To take into account the heat transfer between the sphere and the gas with temperature T lower that of the sphere, the mathematical model was supplemented with the general heat transfer equation [9], also called the heat balance equation [10], which in our case was reduced to the form

$$\rho c_v dT/dt = -P \mathrm{div}\mathbf{V} + \mathrm{div}(\lambda \nabla T) + \Phi_d \qquad (6.8)$$

To relate the gas pressure P to its density ρ of the moving gas (in time and space), in analogy with works [7, 8], the following relationship was used:

$$dP = \gamma P d\rho/\rho. \qquad (6.9)$$

This equation directly follows from the Poisson adiabatic equation $P/\rho^\gamma = \mathrm{const}$, where γ is the adiabatic constant. In our calculations, as for air, we used the value $\gamma = 1.40$.

It was proposed for the same purpose in the monograph [10] that the Clapeyron–Mendeleev equation relating state parameters P, ρ, and T of an ideal gas be used:

$$P = \rho RT/M. \qquad (6.10)$$

However, it should be noted that Eq. (6.10) is valid for the gas in the state of thermodynamic equilibrium. This does not apply to a moving gas with the velocities and temperatures variable in space, for which only a local equilibrium can be considered [11]. Therefore, formula (6.10) can be used to connect the values of P, ρ, and T averaged over time at a given point in space, but not to determine the relations that connect their differentials.

In Eqs. (6.6–6.10) and in other relations below, we use the following notation: R is the universal gas constant, c_v is the specific heat of the gas at constant volume, M is the molar mass, λ and $a = \lambda/(\rho c_v)$ are the heat conductivity and thermal diffusivity, and $\nu = \mu/\rho$ is the kinematic coefficient of the gas viscosity. In calculations, the data on physical properties of the gas were taken to be the same as for air.

Last term Φ_d in Eq. (6.8) describes the dissipation of the mechanical energy of the gas into heat. According to [9, 10], for low velocities of flow and a small gas compression ratio (when we can assume $\mathrm{div}\mathbf{V} \approx 0$), this term can be written in the form

$$\Phi_d = \mu/2(\partial V_i/\partial x_k + \partial V_k/\partial x_i)^2 \qquad (6.11)$$

If the temperatures of the sphere and the gas flowing around it differ only slightly, then changes in coefficients μ and λ can be disregarded, and Eq. (6.8) can be reduced to the form

$$\partial T/\partial t = -\mathbf{V}\nabla T - 0.4T\mathrm{div}\mathbf{V} + a\Delta T + \Phi_d. \qquad (6.12)$$

Taking into account Eq. (6.10) and the data for air, here, we used relations $P/(\rho c_v) = RT/(Mc_v) \approx 0.4T$.

In choosing for simulation of the optimal system of coordinates and the shape of the body in the flow, we took into account the following considerations. If the spherical system of coordinates is used to solve the problem of a gas flow around a sphere in a cylindrical tube with a circular cross section, it is convenient to specify the boundary of the gas flow on the surface of the sphere, but extremely inconvenient to do that on the surface of the tube because the generatrix of the tube surface does not pass through the points of the computational mesh. The inverse situation takes place if we use the cylindrical system of coordinates. Therefore, the following compromise variant [7] was chosen to simplify the problem.

By analogy with the model described in Sect. 5.2, as the body in the gaseous flow, we used a quasi-sphere, i.e., a body of revolution with an octagonal axial section, the sides of which can be obtained by connecting the points of the computational mesh in the cylindrical system of coordinates (Fig. 6.1). This body approximated the sphere of radius $d/2 = 45^{1/2}h \approx 6.71h$ with the center located on the axis of the tube in the middle of the computational domain and could be completely inscribed into this sphere; h is the mesh pitch along coordinates r and z.

To calculate the gas flow around the body in a circular cylindrical tube, we used a uniform computational mesh in the form of a rectangular domain having 26 points along radius r ($i = 0$–25) and 47 points along axial coordinate z ($j = 0$–46).

Equations (6.6), (6.7), and (6.12) in the chosen system of coordinates were written in the finite difference form using the Lax–Wendroff explicit algorithm [13]. In the monograph [13], an approximation of convective terms and the stability of this algorithm were analyzed using the spectral criterion. The same algorithm was previously used by the author of this work in [7, 8]. It is stable when time pitch τ is limited by the Courant– Friedrichs–Lewy condition in the form

$$\tau < h/\left((V + V_s)2^{1/2}\right) \qquad (6.13)$$

where $V = (V_z^2 + V_r^2)^{1/2}$ is the gas velocity modulus, V_z and V_r are the projections of the gas velocity on the coordinate axes, and $V_s = (\gamma P/\rho)^{1/2}$ is the velocity of sound in the gas.

In combination with the Lax–Wendroff algorithm, the approximation of viscous (diffusive) terms of Eqs. (6.7) and (6.12) of the model was carried out according to the explicit algorithm of the first order [13], for which the stability condition in the case of 2D mesh has the form

$$\tau < h^2/(4v), \qquad (6.14)$$

where coefficient a of the gas thermal diffusivity from formula (6.12) must also be used instead of v.

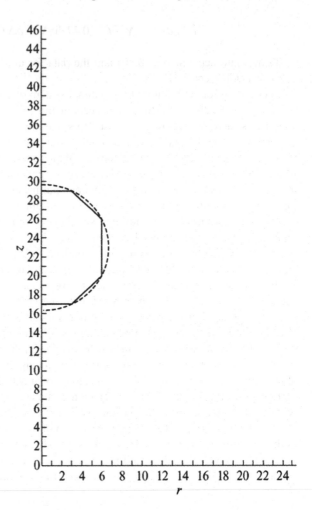

Fig. 6.1 Position half of the axial section of a quasi-sphere relative to the computational mesh; the number of mesh pitches is plotted along the r and z coordinate axes [12]

To ensure the stability of the difference algorithm on the whole, it is necessary that both conditions (6.13) and (6.14) be satisfied simultaneously [13], for which the former condition turned out to be more stringent in the given case.

The difference analogs of differential Eqs. (6.6), (6.7), and (6.12) were supplemented with the corresponding boundary conditions and were solved numerically until the steady-state solution was obtained. In particular, the boundary condition at the solid walls was $\mathbf{V} = 0$ (i.e., zero velocity of the gas). At the cross section of the inlet of the tube and on its wall, a constant gas temperature $T = \text{const} = T_1$ was maintained. The temperature on the surface of the sphere was also kept constant but different, i.e., $T = \text{const} = T_2 = T_1 + \Delta T$, with a small difference $\Delta T \ll T_1$. We calculated the heat flux from the surface of the body with area f to the gaseous flow around it using the calculated field of the gas temperatures near the surface of the body, i.e.,

$$q = dQ/dt = -\lambda \int (n_z \partial T/\partial z + n_r \partial T/\partial r)df. \qquad (6.15)$$

We then calculated coefficient a of the heat transfer from the sphere to the gas and Nusselt number Nu using the formulas

$$a = q/(f\Delta T), \quad \mathrm{Nu} = ad/\lambda \qquad (6.16)$$

The hydrodynamic drag force acting on the body in the direction of the flow was determined using the calculated fields of gas velocities and pressures in the flow by integrating the stresses over surface of the body as follows:

$$F = \int (n_z(-P + 2\mu \partial V_z/\partial z) + n_r\mu(\partial V_z/\partial r + \partial V_r/\partial z))df. \qquad (6.17)$$

Then, we calculated drag coefficient C_d using formula (6.3). Here, $n_z = \cos\theta$ and $n_r = \sin\theta$ are the projections of the unit vector of the outward normal to the surface of the body over the coordinate axes, and θ is the polar angle between the OZ axis and the radius vector **r** drawn from the center of the body to the given point of its surface.

6.1.3   Results of Numerical Experiment

To verify the convergence of the solution of the difference problem to the solution of the differential problem upon a decrease in the mesh pitch, some variants were calculated for two different pitches that differ by a factor of two.

The results obtained for radial profiles of axial velocity $wz[i, j = \mathrm{const}]$ of the gas flowing around a quasi-sphere are shown in Fig. 6.2.

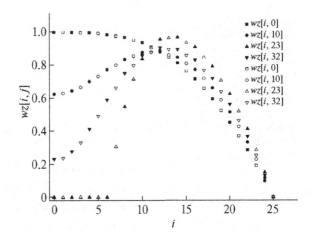

Fig. 6.2 Radial profiles of axial velocity $wz[i, j]$ in the gas flow past the body in the form of a quasi-sphere in a circular cylindrical tube in various cross sections $z_j = \mathrm{const}$ of flow in the Stokes flow regime for $\mathrm{Re} = 0.047$. Light symbols show the results obtained on the mesh with a pitch equals to half of that for which obtained results are shown by dark symbols [12]

For values of i that correspond to odd initial numbers, the results of the calculations were obtained for mesh pitch reduced by half (with dimension $\max(i) \times \max(j) = 50 \times 92$ instead of the initial dimensions 25×46) and are shown by light symbols of the same shape as for values with even numbers i calculated for an initial pitch that is twice as large and shown in Fig. 6.2 by dark symbols. The values of j in both calculations are also corresponding to the initial mesh.

Figure 6.2 shows good agreement between the results of both calculations; the same is observed for radial profiles $t[i, j = \text{const}]$ of the gas temperature, which confirms the convergence of the solution of the difference problem on the mesh. It should also be noted that the velocity profiles shown in Fig. 6.2 are very similar to the same profiles calculated earlier and shown here in Fig. 5.9 and in Fig. 3 in paper [7].

Calculated profiles of the gas temperature are shown in Figs. 6.3, 6.4, 6.5, and 6.6. The axial and radial profiles obtained for the Stokes regime of the flowing a gas

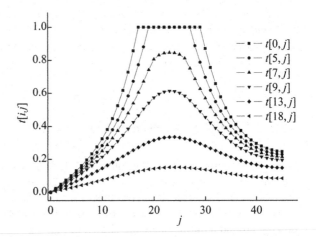

Fig. 6.3 Axial profiles of the relative difference of temperatures $t[i, j]$ of the gas flowing around a body in the form of a quasi-sphere in a circular cylindrical tube at various distances $r_i = \text{const}$ from the flow axis in the Stokes flow regime for $Re = 0.047$ [12]

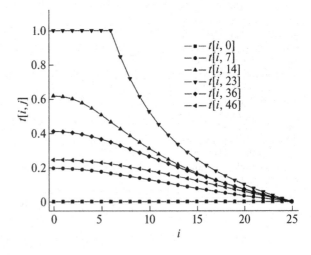

Fig. 6.4 Radial profiles of the relative difference of temperatures $t[i, j]$ of the gas flowing around a body in the form of a quasi-sphere in a circular cylindrical tube in various cross sections $z_j = \text{const}$ of flow in the Stokes flow regime for $Re = 0.047$ [12]

Fig. 6.5 The same as on Fig. 6.3 in the transient regime of flow for Re = 66.3 [12]

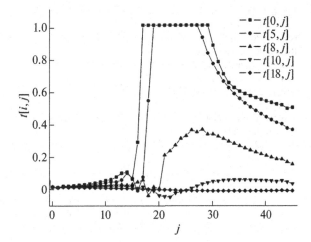

Fig. 6.6 The same as on Fig. 6.4 in the transient regime of flow for Re = 66.3 [12]

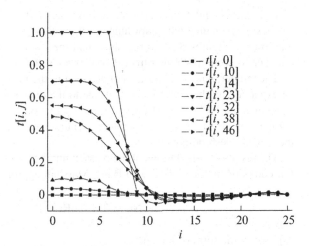

around the body for Re = 0.047 are shown in Figs. 6.3 and 6.4, respectively. The same is shown in Figs. 6.5 and 6.6 for the transient regime with Re = 66.3.

Figure 6.3 shows that, in the Stokes regime, the axial profiles of the gas temperature $(T [r_i, z_j] - T_1)/\Delta T = t[i, j]$ (for i = const) changing with increasing distance from the flow axis along radius r_i still resemble the central symmetry that these profiles would have in the case of heat exchange between the sphere and a stationary gas. Figure 6.5 shows that in the transient regime of flow around the body, these profiles are deformed even more strongly in the direction of the flow, and the abovementioned symmetry almost disappears.

This can be explained (as in the case of a regular sphere) by a change in the nature of the gas flow around the body upon an increase in number Re, in particular, by the appearance of the reverse vortex flow near the stern, which reliably manifested in the gas velocity profiles obtained in [7] and shown in Fig. 5.10.

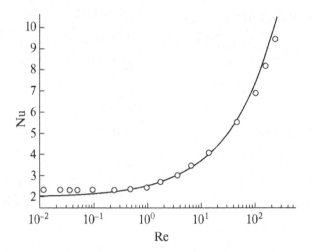

Fig. 6.7 Dependences of Nusselt number Nu of a body on Reynolds number Re: results of calculations using the numerical model of a quasi-sphere in the laminar gas flow in the tube, taking into account the heat transfer (O); solid curve is the calculation by formula (6.18) [12]

Figures 6.4 and 6.6 clearly show how radial (for $j =$ const) gas temperature profiles $t[i, j]$ change with coordinate z_j of the flow cross section in both regimes of the flowing. Figures 6.3 and 6.5 show that the gas temperature at the points of the outlet cross section is higher than at similar points of the inlet cross section of a tube.

The temperature oscillations that are seen in the profiles in Figs. 6.5 and 6.6 can be explained by the numerical effect due to the dispersion of harmonic waves on a difference mesh [11, 13]. However, these oscillations do not prevent from getting a correct idea about the gas temperature variations by averaging the oscillations over the nearest mesh points.

The results of calculations of the Nusselt number are shown in Fig. 6.7 by circles. For comparison, the well-known Ranz–Marshall approximation [14]

$$Nu = 2 + 0.6Re^{1/2}Pr^{1/3} \tag{6.18}$$

is given in the form of the curve.

Figure 6.7 clearly shows that the values of the Nusselt number calculated using the numerical model described above for the body in the gas flowing in the tube are in good agreement with the known data in the whole range of values $0.01 < Re < 240$. Small differences between the values of Nu calculated using the model and formula (6.18) can be explained by the deviation of the shape of the simulating body from the sphere and by the finite number of points of the computational mesh.

It should be noted that the values of number Re were calculated using the gas velocity averaged over the tube inlet cross section.

The results of calculations of drag coefficient C_d are shown in Fig. 6.8 by symbols, and the plots of the known dependences are shown by the curves. Obviously, the calculated values of $C_d(Re)$ for the body in the gas flow are in agreement with Stokes formula (6.4) for $Re < 1$, while for $Re > 5$, these values are in agreement with Klyachko formula (6.5). The small difference between the values of C_d

Fig. 6.8 Dependences of drag coefficient C_d of a body on Reynolds number Re: the results of calculation for a quasi-sphere in the laminar gas flow in the tube using the numerical model ($\circ$); the solid curve is calculated by formula (6.4) for a sphere in the Stokes flow regime with Re < 1; dashed curve is the calculation by formula (6.5) for a transient regime of a flowing with Re > 2 [12]

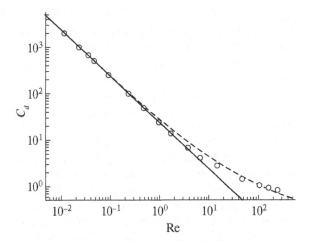

calculated using the proposed model and those calculated by using formulas (6.4) and (6.5) can be explained by the same reasons as for the Nu = Nu(Re) dependence.

The abovementioned good agreement between the results of calculations for the indicated dependences C_d(Re) and Nu(Re) and the known data leads to the conclusion that the deviation of the shape of the body from spherical and the gas flow restriction by wall of a tube weakly affect the values of C_d and Nu. This was established here for the calculating value of the ratio of diameters of the tube and the body $D:d = 25:6$. It is clear that this conclusion is also valid for still more diameters of the tube relative to the diameter of the body ($D:d > 25:6$).

It should be noted that the analogous conclusion concerning the weak effect of the restriction of the gas flow by the channel walls on the value of drag coefficient C_d of the body in the flow was already drawn in [7] for other ratios of $D:d$ of the diameters of the tube and the body (23:6, 29:6, and 35:6).

6.1.4 Hypothesis About Early Crisis of Heat Transfer

The early crisis of drag for a sphere was explained in [7, 8] by the influence of the initially strong turbulence of the incoming flow. High-turbulent viscosity ν_τ creates conditions for the gaseous flow and profiles of its time-averaged velocities which are analogous to the Stokes flow for Re < 1. A small (compared to ν_τ) value of physical viscosity ν, which plays the key role near the surface of the sphere, reduces its drag coefficient C_d in several times.

As noted above (see Sect. 6.1.1), the crisis of drag of a sphere in a strongly turbulent flow must inevitably influence the heat and mass transfer between the sphere and the gas in accordance with the analogy of the transfer of momentum, heat, and mass of the impurity. We tried to estimate this influence in the following way.

Ranz–Marshall Eq. (6.18), which is simplified taking into account the fact that $Pr^{1/3} \approx 1$ for gases, connects parameters $\alpha d/\lambda =$ Nu(Re, Pr) of a convective heat exchange of a spherical particle with gaseous flow.

Due to the crisis of drag, the flow around a sphere is improved, approaching the Stokes flow; in this case, Nu(Re) $\rightarrow$ Nu(0) $= 2$. If the early crisis of drag for droplets occurs in the transient range of numbers Re, then in accordance with Eq. (6.18), $Nu_1 = 7$ for $Re_1 = 70$ and $Nu_2 = 11.5$ for $Re_2 = 250$. This yields ratios Nu_1/Nu $(0) \approx 3.5$ and $Nu_2/Nu(0) \approx 5.8$.

Thus, the intensity of the heat transfer between the drops and the gas due to the early crisis of drag must decrease by 3.5–5.8 times, i.e., in approximately the same way as hydrodynamic drag coefficient C_d of the drops.

This estimate can be verified (we done this later) in the same way as in [8], i.e., by numerical simulation of a strongly turbulent gas flow around a sphere, taking into account the heat transfer between the sphere and the gas. For this purpose, it is necessary to supplement the hydrodynamic model described in [8] with Eq. (6.12), which takes into account the heat exchange between the sphere and the gas.

6.1.5 Conclusions on the Numerical Experiment for a Flow Around a Sphere in a Cylindrical Channel

Thus, the numerical experiment described in Sects. 6.1.2 and 6.1.3 demonstrated that the proposed numerical model of a laminar gaseous flow around a sphere in a cylindrical channel with allowance for heat transfer makes it possible to calculate quite adequately the fields of gas velocities, pressures, and temperatures, drag coefficient for a body in flow, and heat flow and coefficient of heat transfer between a body and gaseous flow. The results of calculations are in good agreement with the known data.

The noted agreement between the results of calculations and the known data for dependences C_d(Re) and Nu(Re) makes it possible to conclude that the gas flow restriction by the channel walls does not significantly influence the values of the indicated quantities in the investigated interval $0.01 < Re < 240$; the ratios of the transverse dimensions of the tube and the body are $D{:}d > 25{:}6$.

In Sect. 6.1.4, we theoretically explained the early crisis of the heat transfer from a sphere to a gas flow with strong turbulence and high-turbulent viscosity using the analogy with the early drag crisis of a sphere in a similar flow. A quantitative estimate of the heat transfer crisis was made, according to which the coefficient of the heat transfer from the sphere to the gas decreases by approximately by three to six times.

6.2 Calculation of Drag Coefficient for a Sphere and Heat Transfer from It to a Gas in Free Laminar Flow and Strongly Turbulent One

An early crisis of drag can occur at high turbulence of gaseous flow around a sphere. To study the influence of a drag crisis on heat transfer between a sphere and a gas, a numerical experiment was carried out in which the free gaseous flow around a sphere with a temperature lower than the sphere temperature was simulated for two cases. The flow was laminar in the first case and highly turbulent in the second case. To take into account turbulence, the kinematic coefficient of turbulent viscosity was introduced, the value of which is much higher (up to 2000 times) than for physical viscosity. The results of calculations show that the early crisis of drag occurs at Reynolds numbers of about 100 and results in considerable (by four to seven times) decrease of the hydrodynamic force and drag coefficient C_d of a sphere. The early crisis of drag is also accompanied by the crisis of heat transfer from a sphere to a gas with a decrease in Nusselt numbers Nu by three to six times.

6.2.1 Introduction to the Problem

In Sect. 6.1 there was already noted the following. In papers [4, 5], it was found on base the experimental results (Fig. 5.15) that, in the highly turbulent two-phase flow with Reynolds number $Re_1 = <V>D\rho/\mu \sim 10^5$ (D – diameter and $<V>$ – mean velocity of gas flow), the value C_d for drops streamlined at $Re = <V>d\rho/\mu \sim 100$ (d – diameter of drops) can decrease by four to seven times compared with the well-known values determined by Klyachko formula (6.5) for the laminar gaseous flow. The same early crisis of drag in the transient region of Reynolds numbers for a single solid sphere was not observed when a sphere was streamlined by the free gaseous flow, but that crisis taken place in the stream flowing through a confuser [6].

Let us note that the experimental points in Fig. 5.15 are near the dashed line that corresponds to the Stokes formula (6.4).

The numerical experiment was not confirmed the assumption about the possible considerable effect of geometry of the gaseous flow on the hydrodynamic drag of the streamlined body [7]. Another explanation of the reason for the early drag crisis of a spherical particle was a hypothesis on the effect of the high turbulence of the gas stream flowing around a sphere. Such turbulence could be enhanced by a confuser compared with a free stream, and as a result, it could become sufficient for the early crisis to occur [6, 7].

This assumption was confirmed by the numerical experiment, in which both the laminar and strongly turbulent free gas flow around a sphere were simulated [8].

As is known, there is an analogy between the phenomena of transfer of heat, mass, and impulse in the inhomogeneous flow. According to this analogy, the profiles of the velocity, impurity concentration, and temperature in different flow cross sections can have some similarities [9, 10]. Taking into account this analogy, we think that the early crisis of drag for a sphere in a strongly turbulent flow should undoubtedly influence with the heat and mass transfer from a sphere to a gas. It was of interest to determine this influence in any way, e.g., by means of numerical simulations of gas flow around a sphere taking into account heat exchange.

This effect was theoretically considered in works [12, 15] and presented in Sect. 6.1.4, where it was shown that the early crisis of drag for a sphere should be accompanied by a decrease in the coefficient of heat transfer from a sphere to gas by 3.5–5.8 times.

The aim of this part of work was the numerical simulation to confirm the influence of the high turbulence of the gas flowing around a sphere (with the occurrence of an early drag crisis) on the heat transfer between a heated sphere and a gas with lower temperature.

6.2.2   Simulation of Free Laminar Gaseous Flow Around a Sphere Taking into Account Their Heat Exchange

The mathematical model of the phenomenon consists of the same differential Eqs. (6.6–6.12) as in Sect. 6.1.

At small temperature differences between sphere and gas stream, changes in coefficients μ and λ can be neglected. As the calculations have shown, the distributions of gas velocities and temperatures are also slightly affected by the term Φ_d in Eqs. (6.8), (6.11), and (6.12). Then, Eq. (6.12) can be reduced to the form

$$\partial T/\partial t = -\mathbf{V}\nabla T - 0.4T\mathrm{div}\mathbf{V} + a\Delta T. \qquad (6.19)$$

Equations (6.6), (6.7), and (6.19) were firstly written in the spherical coordinate system and then represented in the finite–difference form. In the Cartesian two-dimensional coordinate system, the calculation region was shaped as a semiring. In the spherical (polar) coordinate system, it was shaped as a rectangle with sizes of 44 points along radius $r_j = jh$ (layer numbers $j = 10$–53) and 26 points ($i = 0$–26) along an angular coordinate, i.e., polar angle $\theta i = 0 - \pi$ (rad), between polar axis z and the radius vector $\mathbf{r}$ of the given point. The center of symmetry of the region was coinciding with the center of a sphere with radius $R = 10h$, where h is the mesh step along r. The polar axis coincided on the direction with the uniform gas stream far from the sphere.

In the chosen coordinate system, Eqs. (6.6), (6.7), and (6.19) were rewritten into the finite difference form using an explicit Lax–Wendroff scheme [13]. The approximation of the convective terms and stability of the scheme were studied in the same paper [13] using a spectral method. The same scheme was used by the author in earlier works [7, 8]. The scheme is stable when a step τ with respect to time is limited by the Courant–Friedrichs–Lewy condition similar (6.13) in the form

$$\tau < h/\left((V + V_s)2^{1/2}\right), \tag{6.20}$$

where $V = (V_\theta + V_r)^{1/2}$ is the gas velocity module, V_θ and V_r are the projections of gas velocities on coordinate axis, and $V_s = (\gamma P/\rho)^{1/2}$ is the sound speed in gas.

In combination with the Lax–Wendroff scheme, an approximation for the viscous (diffusive) terms in (6.7) and (6.19) was carried out based on an explicit scheme of the first order of accuracy [13]. For this scheme, in the case of a two- dimensional mesh, the condition of the stability can be expressed similar (6.14) in the form

$$\tau < h^2/(4v), \tag{6.21}$$

where the coefficient thermal diffusivity a of the gas from Eq. (6.19) must also be used instead of v.

To provide the stability of a difference scheme as a whole, it is necessary to simultaneously satisfy both conditions (6.20) and (6.21) [13]; in our case the latter is the stricter. The difference analogues of differentials Eqs. (6.6), (6.7), and (6.19) were supplemented by the corresponding boundary conditions and solved numerically as long as a steady-state solution was achieved. In particular, the boundary condition on sphere surface was $\mathbf{V} = 0$, i.e., vanishing gas velocity. At the input boundary of the region ($j = 53$, $i = 13\text{--}26$), the gas density ρ and temperature were kept constant $\rho = \text{const}$ and $T = \text{const} = T_1$. The temperature on the surface of the sphere was also kept constant, but with another value $T = \text{const} = T_2 = T_1 + \Delta T$ with small difference $\Delta T \ll T_1$.

The heat flow from the sphere surface with the area f to the gaseous stream was calculated based on the calculated gas temperature field near the surface of the sphere as follows:

$$q = dQ/dt = -\lambda \int (\partial T/\partial r)df. \tag{6.22}$$

The coefficient α of heat transfer from a sphere to gas and the Nusselt number Nu were then calculated by the formulas (6.16).

The force acting on the sphere was determined using the calculated field of gas velocities and pressures by integrating the mechanical stress on the surface of a sphere as follows:

$$F = \int \left(-P\cos\theta + 3/2v\rho V_\infty \sin^2\theta/R\right)df. \tag{6.23}$$

The drag coefficient of sphere C_d was then calculated from Eq. (6.3).

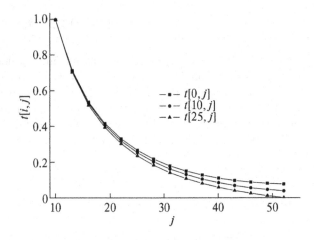

Fig. 6.9 Radial profiles of gas temperature obtained in the Stokes regime of gas flow around a sphere at Re = 0.125 [16]

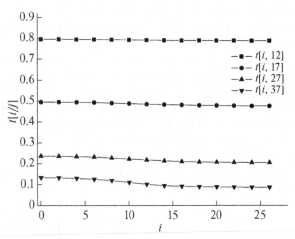

Fig. 6.10 Profiles of gas temperature changing over polar angle θ_i in the Stokes mode of gas flow around a sphere at Re = 0.125 [16]

6.2.3   Results of the Numerical Experiment

The program was constructed by using the aforesaid algorithm and Turbo Pascal to carry out the calculations.

The calculated profiles of gas temperature are shown in Figs. 6.9, 6.10, 6.11, and 6.12. The radial profiles of gas temperature and profiles of its change in respect to polar angle θ_i for the Stokes regime of the laminar gaseous flow around a sphere at Re = 0.125 are shown in Figs. 6.9 and 6.10, respectively. The same profiles for the transient mode at Re = 128 are shown in Figs. 6.11 and 6.12.

It is obviously from Fig. 6.9 that, in the Stokes mode, the radial profiles of gas temperature ($T[\theta_i, r_j] - T_1)/\Delta T = t[i, j]$ (at $i = $ const) are slightly different for different polar angles θ_i, and it reminds about central symmetry, which these profiles would have at heat exchange between a sphere and immobile gas. It is obvious in Fig. 6.11 (in comparison with Fig. 6.9) that, at the flow around a sphere in the

Fig. 6.11 Same as in Fig. 6.9 but in the transient regime of gaseous flow around a sphere at Re = 128 [16]

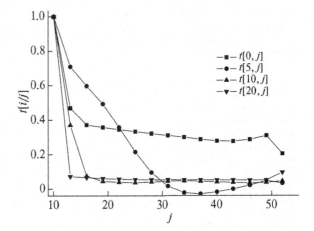

Fig. 6.12 Same as in Fig. 6.10 but in the transient regime of gaseous flow around a sphere at Re = 128 [16]

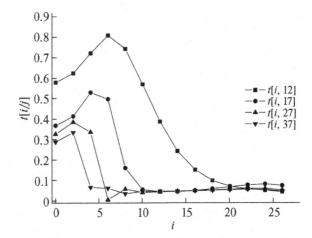

transient regime, the radial profiles are strongly distorted in the flow direction and the abovementioned symmetry is already absent. This fact is caused by a change in the behavior of gas flow around a sphere with increasing the number Re, in particular by the occurrence of the reciprocal vortex flow near the stern region of a sphere; this flow was the very pronounced in the gas velocity profiles given in [8].

It is obvious from Fig. 6.10 that, in the Stokes regime of a flow, the gas temperature $t[i, j]$ at j = const changes slightly with respect to polar angle θ_i. Vice versa, as it can be seen from Fig. 6.12, in the transient regime of a flow, the gas temperature $t[i, j]$ (at j = const) changes with respect to angle θ_i, rather substantially at different distances r_j from the sphere surface.

The Nu numbers for a sphere streamlined by a free gas flow were calculated using the aforesaid numerical model. The results of calculations for the Nusselt numbers are shown in Fig. 6.13 by circles. For comparison the known Ranz–Marshall approximation [14] in the form (6.18) is shown by a curve. Obviously, they are in

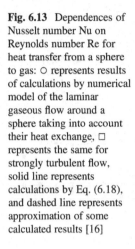
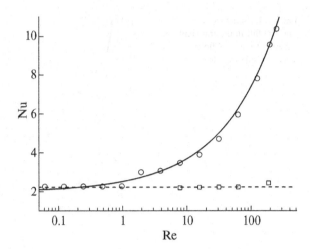

Fig. 6.13 Dependences of Nusselt number Nu on Reynolds number Re for heat transfer from a sphere to gas: ○ represents results of calculations by numerical model of the laminar gaseous flow around a sphere taking into account their heat exchange, □ represents the same for strongly turbulent flow, solid line represents calculations by Eq. (6.18), and dashed line represents approximation of some calculated results [16]

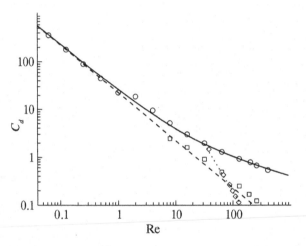

Fig. 6.14 Dependences of drag coefficient C_d of sphere on the Reynolds number in logarithmic scale: ○ represents the results of calculations by numerical model of laminar gaseous flow around a sphere taking into account heat exchange; □ represents the same for strongly turbulent flow; dashed line represents calculations by Eq. (6.4); solid line represents calculations by Eq. (6.5); ◇ represents experiment [4] with water drops in spray of nozzle at pressure $P = 5$ atm; dotted line represents approximation of experimental data by formula $C_d = 2000/Re^2$ [16]

good agreement with the known data over a whole studied range $0.06 < Re < 400$. A small difference between the Nu numbers calculated using the model and Eq. (6.18) at $Re = 1-2$ can be accounted for by the transition from the Stokes to transient regime, as well as by the limitation of a number of computational grid nods.

The results of calculations for drag coefficient C_d of a sphere in comparison with the known dependences are shown in Figs. 6.14 and 6.15 (in different scales) by circles and solid lines, respectively. The experimental data of water drops in spray of a nozzle are shown by rhombuses. Obviously, calculated at $Re < 1$ values of C_d for a

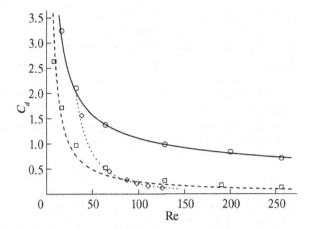

Fig. 6.15 Same as in Fig. 6.14 but on a linear scale [16]

sphere streamlined by a gas flow are in agreement with the Stokes formula (6.4), and values calculated at Re > 2 are in agreement with the Klyachko formula (6.5). The small difference between values C_d calculated by using a suggested model and Eq. (6.4) or (6.5) at Re = 1–2 (see Fig. 6.14) can be explained by the same reasons as for the dependence Nu = Nu(Re).

6.2.4 Simulation of Strongly Turbulent Free Gaseous Flow Around a Sphere Taking into Account Their Heat Exchange

When simulating a turbulent flow around a sphere, the following considerations were used. If a strongly turbulent gas stream flows around a fixed sphere in the transitional range (Re ~ 10–10^2), for example, a round jet with a diameter D, then it can be described by the Reynolds number $Re_1 = <V>D\rho/\mu \sim 10^5$ and by the kinematic coefficient of turbulent viscosity ν_τ, which can be taken constant within the stream and determined by the formula

$$\nu_\tau = \sigma(J/\rho)^{1/2} = \sigma(\pi/4)^{1/2}Re_1\nu = \text{const} \approx 0.02\ Re_1\nu \sim 2\cdot 10^3\nu \gg \nu, \quad (6.24)$$

Here, $J = \pi/4D^2\rho < V>^2 = \text{const}$ is the momentum flux of the gaseous stream, $<V>$ is its mean velocity over cross section, and $\sigma \approx 0.021$ is the empirical constant [10]. Let us note that, according to formula (6.24), in this case, the kinematic coefficient of turbulent viscosity ν_τ is much more than the similar coefficient of usual physical viscosity ν.

According to the theory of near-wall turbulence, a turbulent boundary layer is formed near the surface of the streamlined body; pulse transfer in the layer is determined by the sum action of turbulent and physical viscosities $\nu_\Sigma = \nu_\tau(y) + \nu$

[10]. The first term $\nu_\tau = (0.4y)^2 |\partial V_\theta/\partial y|$ changes proportionally to square of the distance $y = r - R$ from sphere surface, and at small thickness ($\delta \ll R$) of the boundary layer increases drastically from zero to maximum value determined by formula (6.24) for a flow far from the streamlined body.

The Reynolds equations for time-averaged variables, velocity, density, and pressure, which describe a quasi-stationary flow in the turbulent boundary layer, have the same form as Eqs. (6.6), (6.7), and (6.9), but differ by the substitution of ν_Σ for ν.

In the described numerical experiment, the kinematic coefficient of turbulent viscosity was approximated by the function

$$\nu_\tau(y) = \nu_\tau(\infty)(1 - R/r)^2 = 2000\nu(1 - R/r)^2, \qquad (6.25)$$

according to which $\nu_\tau \to 0$ at $y = (r - R) \to 0$ and $\nu_\tau \to const = 2000\nu$ at $y \to \infty$, that corresponds to the representations of the theory of near-wall turbulence.

To calculate the gas temperature in the high-turbulent stream, formula (6.19) was used, the same as for the laminar flow.

A pressure change in the cross section of the stream is usually neglected in the high-turbulent flow [10]. If we assume that, in our case, it is valid until the boundary layer is not separate from the surface of the body, the first term in the integrand in Eq. (6.23) can be neglected.

The radial profiles of gas temperature and profiles of its change over the polar angle θ_i are shown in Figs. 6.16 and 6.17, respectively. The profiles were calculated using numerical simulation of high-turbulent flow around the sphere in the transient range at Re = 128. Their comparison with the profiles in Figs. 6.9 and 6.10 show their noticeable similarity (so as for velocity fields [8]), which is caused by the influence of high-turbulent viscosity of the gas stream. Vice versa, the considerable difference between the profiles is evident in Figs. 6.16, 6.17, 6.11, and 6.12.

The results of calculations of the drag coefficient C_d of the sphere obtained at numerical simulations of the high-turbulent flow around a sphere (Re$_1 \sim 10^5$) are

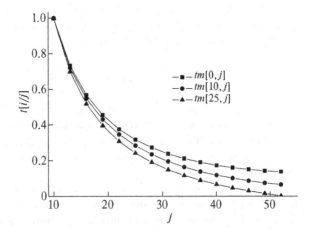

Fig. 6.16 Calculated radial profiles of gas temperature change for strongly turbulent gas flow around a sphere in the transient range at Re = 128 [16]

Fig. 6.17 Calculated profiles of gas temperature change over polar angle θ_i for strongly turbulent gaseous flow around a sphere in the transient range at Re = 128 [16]

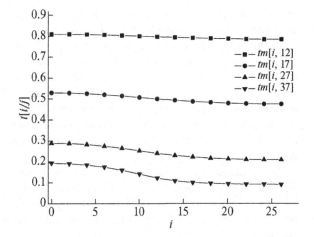

shown in Figs. 6.14 and 6.15 by squares located close to a dashed line that corresponds to the Stokes formula Eq. (6.4). The experimental data [4] obtained for water drops in a spray of nozzle at $P = 5$ atm are also shown by rhomb symbols in the same figures. It is obvious from Fig. 6.15 that, at Re > 50, the results of experiments and calculations for the high-turbulent flow around a sphere are in better agreement than for the laminar flow. This fact confirms the correctness of the suggested model ideas of a mechanism for the early crisis of the body drag in the highly turbulent flow.

The great difference between the results of experiment and calculation at Re < 50 can be accounted for by the fact that, under the experimental conditions, the level of turbulence of a gas flow was still insufficient to lead to the early crisis of drag for droplets at this values range of Re, which corresponds to smaller distances from the nozzle [4].

6.2.5 Confirmation of Hypothesis About Early Crisis of Heat Transfer

In papers [7, 8], the early crisis of drag for sphere is accounted for by the influence of initial high turbulence of the incoming gaseous flow. The high-turbulent viscosity ν_τ of gas produces the flowing conditions and time-averaged profiles of gas velocities similar to Stokes flow at Re < 1. The low physical viscosity ν (compared with ν_τ), which has a dominant role near the sphere surface, decreases the drag coefficient C_d of a sphere by several times.

As was mentioned in Sect. 6.2.1, taking into account the analogy between the phenomena of the transfer of impulse, heat, and mass, the crisis of drag for a sphere in strongly turbulent flow should certainly affect its heat and mass exchange with gas. In work [15], author tried to estimate this influence. The obtained estimates were presented in work [12] and Sect. 6.1.4 as follows.

In the critical region of Reynolds numbers, the flow around a spherical drop improves approaching the Stokes flow, while $Nu(Re) \rightarrow Nu(0) = 2$. In the transient range of Re numbers, where the early drag crisis of drops occurs, $Nu_1 = 7$ at $Re_1 = 70$ and $Nu_2 = 11.5$ at $Re_2 = 250$ according to Eq. (6.18). Then, we have relationships $Nu_1/Nu(0) \approx 3.5$ and $Nu_2/Nu(0) \approx 5.8$.

Thus, the intensity of the heat exchange between a droplet and gas at early drag crisis should decrease in 3.5–5.8 times, almost the same as the hydrodynamic drag coefficient C_d of a droplet.

This estimate was check as in [8] by means of the numerical simulations of the high-turbulent gas flow around a sphere taking into account the heat exchange between a sphere and gas. To do this, the hydrodynamic model described in [8] was supplemented with Eq. (6.19) taking into account the heat exchange between a sphere and gas.

The results of calculations of Nusselt number as a function of Reynolds number Re by using the aforementioned model of the high-turbulent gas flow around a sphere taking into account the heat exchange between a sphere and gas are shown in Fig. 6.13 by the square symbols. Obviously, in the transient range of Reynolds numbers Re = 8–200, the Nu numbers differ slightly from the values typical for the Stokes regime of the laminar flow around a sphere and are approximately equal to 2.2. This circumstance is shown in Fig. 6.13 by a dashed line.

6.2.6 Conclusions on the Results of Numerical Experiment

Thus, the numerical experiment described in Sect. 6.2 showed the following. First, the suggested algorithm of calculation of the laminar gaseous flow around a sphere taking into account the heat exchange allowed us to satisfactorily calculate the fields of gas velocities, pressures, and temperature, also the values of force and coefficient of drag for a sphere, and also heat flow and coefficient of heat transfer from a sphere to gas flow. The results of calculations are in good agreement with the known data.

Second, a combination of the suggested algorithm with elements of the near-wall turbulence theory allowed to simulate the strongly turbulent flow around a sphere as well as to calculate the fields of gas velocities and temperatures. Also, as a result, the drag coefficient C_d of the sphere was calculated (in accordance with the experimental data for early crisis of drag), and the Nusselt number Nu that characterizes the heat emission of a sphere was determined. Furthermore, the values of Nu turned out to be several (3–6) times lower compared with Nu for the laminar flow around a sphere in the transient regime. This fact corroborates the validity of the assumption suggested in works [12, 15] about that the early crisis of drag can be accompanied by the early crisis of heat transfer.

The results presented in Sects. 6.1 and 6.2 were previously published in articles [12, 16].

6.3   Numerical Simulation and Calculation of Heat Exchange a Separate Droplet with a Gas Stream

This problem arises, for example, in connection with the calculation of the process of spray granulation of polymers from their melt, in particular polyethylene wax, when it is required to determine how quickly the drop cools and solidifies. Inner convection can be ignored due to the high viscosity of the melt. As noted in Chap. 1, when modeling and calculating complex processes in a heterogeneous system, its hydrodynamic model being as a basis must be supplemented by models of kinetics of elementary heat and mass transfer acts that represent solutions to external, internal, or mixed task of interphase heat and mass exchange [1].

6.3.1   Mathematical Model and Numerical Algorithm

When solving the heat transfer problem (without allowance for evaporation) of a separate drop, at the surface temperature $T_s = T(R, \tau)$ and with the surrounding gas having the temperature $T_g(\tau)$, we use the heat equation

$$\rho \cdot c \cdot \partial T / \partial \tau = \operatorname{div}(\lambda \cdot \operatorname{grad} T) \qquad (6.26)$$

with a boundary condition of the third kind (mixed problem of heat exchange)

$$-\lambda \cdot \partial T(R, \tau) / \partial r = a \cdot (T_s - T_g). \qquad (6.27)$$

under the initial condition

$$T(r, 0) = T_0. \qquad (6.28)$$

When the thermal conductivity ($\lambda = \text{const}$) of the droplets substance is constant, the thermal diffusivity $a = \lambda/(\rho \cdot c)$ can be introduced, and then Eq. (6.26) takes a simpler form

$$\partial T / \partial \tau = a \cdot \Delta T. \qquad (6.29)$$

Note that, unlike the analytical solution of the problem (see Sect. 1.5), this transition is of no fundamental importance for the numerical method, and it is not much harder to solve the problem when the quantities λ, c, ρ are changing together with the temperature of droplet, and also if the gas temperature $T_g(\tau)$ and its rate relatively droplet, together with heat transfer coefficient a are changing during heat exchange. But in this example, for simplicity, we will not take into account these changes, assuming λ, c, ρ, $T_g = \text{const}$.

Just as in the analytic solution it is necessary to go over to dimensionless variables $x = r/R$, $t = a \cdot \tau\ R^2 = $ Fo is the Fourier number, $U(x, t) = (T - T_g)/(T_0 - T_g)$ with using the dimensionless parameter Bi $= a \cdot R/\lambda$ – the Bio number.

Taking into account the spherical symmetry of the droplet, we disclose the form of the Laplace operator Δ, and relatively dimensionless variables, instead of (6.27), (6.28), and (6.29), obtain the equations

$$\frac{\partial U}{\partial t} = \frac{1}{x^2} \cdot \frac{\partial}{\partial x}\left(x^2 \cdot \frac{\partial U}{\partial x}\right) \qquad (6.30)$$

$$-\frac{\partial U(1,t)}{\partial x} = \mathrm{Bi} \cdot U(1,t) \qquad (6.31)$$

$$U(x,0) = 1 \qquad (6.32)$$

We supplement it with a completely natural left boundary condition (at the center of the droplet)

$$\frac{\partial U(0,t)}{\partial x} = 0 \qquad (6.33)$$

Passing to the discrete values of the variables – $t^n = n \cdot ht$, $n = 0, 1, 2, ...;$ $x_m = m \cdot hx = m/M$, $m = 0, 1, 2, ..., M$; $U_m^n = U(x_m, t^n)$ – and to the finite differences in the representation of the derivatives, we use the implicit scheme of the triangle [11], which has unconditional numerical stability. In this case we obtain the following difference equations:

$$\frac{U_m^{n+1} - U_m^n}{ht} = \frac{1}{x_m^2} \cdot \frac{1}{hx} \cdot \left(x_{m+1/2}^2 \cdot \frac{U_{m+1}^{n+1} - U_m^{n+1}}{hx} - x_{m-1/2}^2 \cdot \frac{U_m^{n+1} - U_{m-1}^{n+1}}{hx}\right) \quad (6.34)$$

$$\frac{U_1^{n+1} - U_0^{n+1}}{hx} = 0, \qquad -\frac{U_M^{n+1} - U_{M-1}^{n+1}}{hx} = \mathrm{Bi} \cdot U_M^{n+1} \qquad (6.35)$$

$$U_m^0 = 1 \qquad (6.36)$$

These algebraic equations are reduced to a system of a special kind

$$-b_0 \cdot U_0^{n+1} + c_0 \cdot U_1^{n+1} = d_0$$
$$a_m \cdot U_{m-1}^{n+1} - b_m \cdot U_m^{n+1} + c_m \cdot U_{m+1}^{n+1} = d_m, \qquad (6.37)$$
$$a_M \cdot U_{M-1}^{n+1} - b_M \cdot U_M^{n+1} = d_M$$

which has a tridiagonal matrix and can be solved by the sweep method [11].

Fig. 6.18 The droplet temperature profiles at Bi = 0.1; t = 0.1; 0.4; 0.8; 1.2; 1.6; 2.0; 2.4; 2.8 – from top to bottom

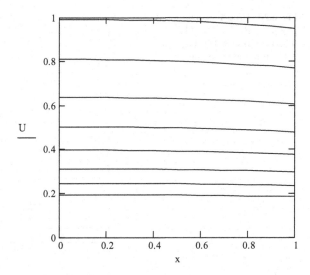

Fig. 6.19 The droplet temperature profiles at Bi = 0.3; t = 0.1; 0.8; 1.6; 2.4; 3.2; 4.0; 4.8; 5.6 – from top to bottom

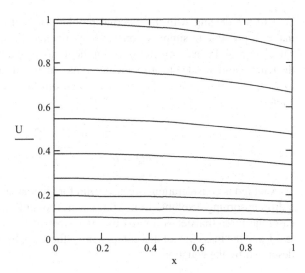

6.3.2 *Calculation Results*

Omitting here the details of the programming of the algorithm proposed above, let us consider the results of calculating the cooling of a heated drop in a colder gas stream, obtained for the values of the number Bi = 0.1; 0.3; 1 and presented in Figs. 6.18, 6.19, and 6.20.

Obviously, for not too intense heat transfer (Bi < 0.3), which is true for water droplets in the jet of the nozzle (see Sect. 1.5), during the whole process of cooling the droplet, the temperature profiles inside it can be considered practically uniform,

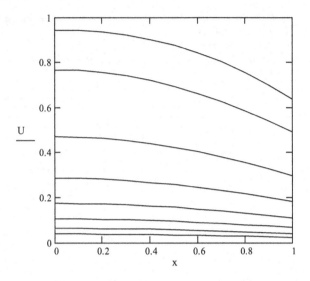

Fig. 6.20 Drop temperature profiles at $Bi = 1$; $t = 0.1$; 0.2; 0.4; 0.6; 0.8; 1.0; 1.2; 1.4 – from top to bottom

i.e., we are dealing with the external problem of heat transfer. This circumstance makes it possible to simplify considerably mathematical description and numerical calculation of the heat transfer between the droplets and the gas flow, assuming that the temperature inside the drop, including its center and surface, is everywhere equal to $T_d(t)$, and using for it and the heat flux from the gas to the droplet the simple equations:

$$c \cdot m_d \cdot dT_d = dQ \tag{6.38}$$

$$dQ = \alpha \cdot \left(T_d - T_g\right) \cdot S \cdot dt. \tag{6.39}$$

Analogous considerations are also valid for the mass exchange between a gas and a drop absorbing a soluble in its component (possible with chemisorption). In this case, the mass transfer is limited by the gaseous phase (the external mass transfer task), and the mass flux of this component to the drop, by analogy with (6.39), can be described by the equation

$$dM = \alpha_D \cdot \left(\rho_g - \rho_s\right) \cdot S \cdot dt \cong \alpha_D \cdot \rho_g \cdot S \cdot dt. \tag{6.40}$$

In the last approximate equality, it is additionally taken into account that, under the given conditions, the density of the component which is absorbing by drop outside drop but near its surface is small $\rho_s \ll \rho_g$. Here $\alpha_D = \text{Nu} \cdot D/d$ is the mass transfer coefficient, $d = 2R$ is the diameter of droplet, and D is the diffusion coefficient of the component in the gas.

6.3.3   Conclusion on Modeling of Elementary Acts Kinetics of Heat and Mass Transfer

In many practically important cases, interphase heat and mass transfer are limited to its transfer in the gas phase, which makes it possible to simplify considerably the mathematical description of the transfer phenomenon and the calculation of its kinetics.

References

1. Brounshtein, B. I., & Fishbein, G. A. (1977). *Fluid dynamics, mass and heat transfer in dispersed systems*. Leningrad: Khimiya.
2. Torobin, L. B., & Gauvin, W. H. (1959). *Canadian Journal of Chemical Engineering, 37*(4), 129.
3. Schlichting, H. (1955). *Boundary-layer theory*. New York: McGraw-Hill. Nauka, Moscow, 1974).
4. Simakov, N. N. (2004). Crisis of Hydrodynamic Drag of Drops in the Two-Phase Turbulent Flow of a Spray Produced by a Mechanical Nozzle at Transition Reynolds Numbers. *Technical Physics, 49*, 188.
5. Simakov, N. N., & Simakov, A. N. (2005). Anomaly of gas drag force on liquid droplets in a turbulent two-phase flow produced by a mechanical jet sprayer at intermediate Reynolds numbers. *Journal of Applied Physics, 97*, 114901.
6. Simakov, N. N. (2010). Experimental Verification of the Early Crisis of Drag Using a Single Sphere. *Technical Physics, 55*, 913.
7. Simakov, N. N. (2011). Effect of the gas flow geometry and turbulence on the hydrodynamic drag of a body in the flow, Technical Physics. *Technical Physics, 56*, 1562.
8. Simakov, N. N. (2013). Calculation of the flow about a sphere and the drag of the sphere under laminar and strongly turbulent conditions. *Technical Physics, 58*, 481.
9. Landau, L. D., & Lifshitz, E. M. (1988). *Course of theoretical physics* (Fluid mechanics) (Vol. 6). Moscow: Nauka. Pergamon, New York, 1987).
10. Loitsyanskii, L. G. (1978). *Mechanics of liquids and gases*. Moscow: Nauka. Pergamon, Oxford, 1966).
11. Fedorenko, R. P. (1994). *Introduction to computational physics*. Moscow: Moscow Institute of Physics and Technology.
12. Simakov, N. N. (2016). Calculation of the Drag and Heat Transfer from a Sphere in the Gas Flow in a Cylindrical Channel. *Technical Physics, 61*, 1312.
13. Potter, D. (1973). *Computational physics*. New York: Wiley. Mir, Moscow, 1975.
14. Ranz, W. E., & Marshall, W. R. (1952). *Chemical Engineering Progress, 48*(5), 173.
15. Simakov, N. N. (2014). Early crisis of ball resistance in a highly turbulent flow and its effect on heat and mass transfer of a ball with a gas. In V. N. Blinichev (Ed.), *Proceedings of the international scientific–technical conference on problems of resource-and energy-saving technologies in industry and agro-Industrial complex, Ivanovo, 2014* (Vol. 2, p. 389). Ivanovo: Ivanovo State University of Chemistry and Technology.
16. Simakov, N. N. (2016). Calculations of the Flow Resistance and Heat Emission of a Sphere in the Laminar and High-turbulent Gas Flows. *Technical Physics, 61*, 1806.

Chapter 7
Numerical Simulation of Two-Phase Flow Produced by a Nozzle Taking into Account the Early Crisis of Drag for Droplets and Interphase Mass Transfer

A numerical experiment on the simulation of the two-phase flow formed during spraying of a liquid in a gas by a nozzle has been described. The distinguishing feature of the mathematical model is that it employs the differential equations describing the nonstationary flow of a compressible fluid as the initial equations. To transit to their difference analog, the known Lax–Wendroff algorithm has been used. The same numerical model was used to calculate the interfacial mass transfer in spray flow. For this purpose the basis differential equations of the proposed model have been supplemented by the equation of impurity mass transfer from gas to droplets.

7.1 Introduction: Peculiarities of the Two-Phase Flow of the Spray

For the intensification of heat and mass transfer processes such as burning liquid fuels, drying and granulating polymers, and wet cleaning of air from dust and harmful gaseous impurities, by increasing the interphase surface, the spraying of liquid in a gas is used, for example, with use injectors.

The basis for calculating such processes is the knowledge of hydrodynamic structure of created spray and force of interaction of droplet with gas and also the conception of elementary act of heat and/or mass transfer at the level of individual droplet. So far, methods of satisfactory calculation of spraying processes were not sufficiently developed, which is the reason of the actuality of this work.

Two basic approaches are known for mathematical modeling of two-phase flows: the method of interpenetrating continua [1] and the theory of turbulent jets [2]. In the first of these, each of the phases is considered as continuously distributed in space with a variable (averaged in small volume) density, and the phase velocities are assumed to be different. In the second, it is assumed that the concentration of the

© Springer Nature Switzerland AG 2020
N. N. Simakov, *Liquid Spray from Nozzles*, Innovation and Discovery in Russian Science and Engineering, https://doi.org/10.1007/978-3-030-12446-5_7

dispersed phase is small, the phase velocities are approximately the same, but the turbulence of the gaseous phase flow is taken into account. Each of these approaches, taken separately, does not take into account the important features of the spray flow, including the following.

It has been established experimentally that the gas flow in the spray is a turbulent jet [3, 4]. This jet is formed at the root of the spray due to the interaction between phases and subsequently evolves as if autonomously from the droplet flow. This jet differs from the one-phase flow in the structure and the type of turbulent friction. In particular, dimensionless profiles of the axial velocity of the gas turn out to be slightly different (more gently sloping) than in a one-phase jet. It has been established that the velocities of the phases at each point of the flow are different, and the gas pressure differences on the order of 1–10 Pa exist along the spray axis and radius.

In addition, the substantial peculiarity in the interaction between the phases (early drag crisis) has been detected. This should be taken into account in calculating the spray and can be explained as follows.

In processes involving the spraying of a liquid, droplets with average diameter d on the order of 10^{-4} m are formed. With this size and a large difference in the dynamic viscosities of the liquid in droplets and of the gas flowing past them (by about 60 times in the case of water and air), the deformation of droplets and the internal flow of the liquid in them can be disregarded, treating droplets as hard spheres.

As known, e.g., from works [5, 6], the hydrodynamic drag force acting a droplet in a gas flow can be calculated using the formulas (6.3) and (6.4) or (6.5). For the convenience of readers, we again give these formulas here.

$$F = C_d S \rho V^2 / 2. \tag{7.1}$$

$$C_d = 24/\text{Re} \tag{7.2}$$

$$C_d = 24/\text{Re} + 4/\text{Re}^{1/3} \tag{7.3}$$

It was shown in [3, 4] using experimental data (Fig. 7.1) that the value of C_d for droplets in a strongly turbulent flow with Re $\approx$ 100 is smaller by a factor of 4 to 7 compared to the well-known values determined by Klyachko formula (7.3). An analogous early drag crisis was also observed for a solitary hard sphere in a gas jet flowing through a confuser [7]. It should be noted that experimental points shown in Fig. 7.1 for $z > 0.15$ m are close to the values that correspond to Stokes formula (7.2).

As a reason explaining the early crisis of drag for a spherical particle, the hypothesis was put forth in [7] concerning the effect of strong turbulence of the gas flow, which was intensified by the confuser still further (as compared to a free jet) so that it became sufficient for the emergence of an early crisis of drag for a solitary hard sphere. This hypothesis was confirmed by a numerical experiment on a sphere in a free gas flow (both laminar and strongly turbulent) [9]. The above arguments lead to the conclusion that it is expedient to use a combination of the

Fig. 7.1 Dependence of drag coefficient C_d of droplets at distance z from the nozzle: (■) experimental results [3] with water droplets on the spray axis under water pressure $P_1 = 5 \times 10^5$ Pa at the sprayer; (●) same at points of the spray boundaries; solid curve is an approximation of experimental results at spray axis by formula (7.15); dashed curve is calculated by formula (7.3) using experimental data from [3] for velocities of gas and droplets and their sizes [8]

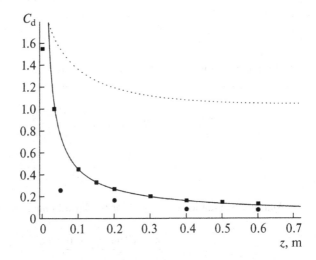

two abovementioned approaches (method of interpenetrating continuums [1] and the theory of turbulent jets [2]) in mathematical and numerical simulation of a spray as a two-phase flow, taking into account all its peculiarities to describe the motion of both phases by uniform method. An analogous idea was used earlier in [4, 10], but to a limited extent (on a domain of calculation with a smaller number of points, 26×26, and only for a free spray unbounded by the apparatus walls).

7.2   Numerical Simulation and Calculation of Hydrodynamics of Two-Phase Flow Produced by a Nozzle

The radial and axial velocity profiles of the droplets and gas in the free spray flow and in the two-phase stream through a cylindrical apparatus have been calculated and represented taking into account the early crisis of drag for droplets and peculiarities of turbulent friction in the gas, which was detected in previous experiments. A comparison of the results of calculations based on this model with experimental data has demonstrated their agreement.

7.2.1   Mathematical Model of a Free Spray

A turbulent flow of the gas phase in the cylindrical system of coordinate can be described by the time-dependent continuity equation

$$\frac{\partial \rho}{\partial t} + \frac{\partial (\rho w_z)}{\partial z} + \frac{1}{\rho}\frac{\partial (r\rho w_r)}{\partial r} = 0 \qquad (7.4)$$

and by the equations of motion for the time-averaged axial and radial gas velocity components $w_z = w_z(r, z)$, $w_r = w_r(r, z)$ as follows:

$$\frac{\partial w_z}{\partial t} + w_z \frac{\partial w_z}{\partial z} + w_r \frac{\partial w_z}{\partial r} = \frac{1}{\rho r}\frac{\partial (r\tau)}{\partial r} - \frac{1}{\rho}\frac{\partial P}{\partial z} + \frac{f_z}{\rho} \qquad (7.5)$$

$$\frac{\partial w_r}{\partial t} + w_z \frac{\partial w_r}{\partial z} + w_r \frac{\partial w_r}{\partial r} = -\frac{1}{\rho}\frac{\partial P}{\partial r} + \frac{f_r}{\rho} \qquad (7.6)$$

Analogous equations are used for the liquid phase (by using subscript "l"):

$$\frac{\partial \alpha}{\partial t} + \frac{\partial (\alpha u_z)}{\partial z} + \frac{1}{r}\frac{\partial (r\alpha u_r)}{\partial r} = 0, \qquad (7.7)$$

$$\frac{\partial u_z}{\partial t} + u_z \frac{\partial u_z}{\partial z} + u_r \frac{\partial u_z}{\partial r} = -\frac{f_z}{\alpha \rho_1}, \qquad (7.8)$$

$$\frac{\partial u_r}{\partial t} + u_z \frac{\partial u_r}{\partial z} + u_r \frac{\partial u_r}{\partial r} = -\frac{f_r}{\alpha \rho_1} \qquad (7.9)$$

Here, ρ and P are the gas density and pressure, τ is the turbulent drag stress in the gas phase, f_r and f_z are the volume densities of the interfacial interaction forces, α is the relative volume of the liquid, ρ_1 is the physical density of droplets, and u_z and u_r are the axial and radial velocity components of the liquid.

System of Eqs. (7.4–7.9) is not closed since it contains nine unknown functions, i.e., w_z, w_r, ρ, P, α, u_z, u_r, f_z, and f_r. The closure of the system can be performed as follows.

First, one can assume that the gas flow in the spray is adiabatic, and we can use the equation of the Poisson adiabat:

$$\frac{P}{\rho^\gamma} = \text{const} = \frac{P_0}{\rho_0^\gamma} \qquad (7.10)$$

$$dP = \gamma \frac{P}{\rho} d\rho = c^2 d\rho = c_0^2 \left(\frac{\rho}{\rho_0}\right)^{\gamma-1} d\rho \qquad (7.11)$$

where c is the velocity of sound in the gas, γ is the adiabatic exponent, and subscript "0" marks the initial values of the quantities.

Substituting the last term from (7.11) into expressions (7.5) and (7.6), we can eliminate gas pressure P from system of Eqs. (7.4–7.9).

Second, based on experimental data, the following expression was obtained in [4] for turbulent friction stress τ in a gas:

$$\tau = -\rho \frac{rZ}{2\zeta^2} \left(\frac{\partial w_z}{\partial r} \right)^2 \qquad (7.12)$$

where $Z = z + Z_0$ is the axial coordinate measured from the gas jet pole, z is the same but that measured from the sprayer, Z_0 is the distance from the pole to the sprayer, and $\zeta = \mathrm{const}(P_1)^{1/2}$. For excess water pressure $P_1 = 5 \times 10^5$ Pa at the sprayer, values $Z_0 = 390$ mm and $\zeta = 11.9$ were obtained.

It should be noted that representation (7.12) differs from a following analogous representation for a one-phase turbulent gas jet in accordance with the new Prandtl hypothesis:

$$\tau = \rho l^2 \left(\frac{\partial w_z}{\partial r} \right)^2 = \rho_\mu \frac{\partial w_z}{\partial r} \qquad (7.13)$$

where l is the mixing length. For a circular jet, the kinematic coefficient of turbulent viscosity $\nu_t = \mathrm{const}$ is an empirical constant [11]. It should be noted that, by substituting expression (7.12) into (7.5), we can disregard the dependence of ρ on r in the subsonic flow for simplifying the latter equation.

Third, the drag exerted by the gas to an individual droplet is usually expressed by formula (7.1), where $\mathbf{V} = \mathbf{U} - \mathbf{W}$ is the relative velocity of the droplet in the gas. Then we can write the following expression for the components of the volume density of interfacial interaction forces:

$$f_{z,r} = F_{z,r} n = \frac{F_{z,r} \cdot \alpha \cdot \rho_1}{m_d} \qquad (7.14)$$

where n is the numerical concentration of droplets and m_d is the mass of an individual droplet.

As noted above, the values of drag coefficient C_d for droplets in the spray and its dependence on Reynolds number $\mathrm{Re} = \rho V d / \mu$ are anomalous [3, 4]. Near sprayer the values of C_d approximately correspond to the dependences which are well known from the literature (e.g., Klyachko formula (7.3)) and sharply decrease (by four to seven times) with increasing distance from the sprayer (see Fig. 7.1).

It can be seen from Fig. 7.1 that the decrease in C_d with increasing distance z from the sprayer along the its axis is successfully approximated by the dependence

$$C_d(0, z) = 0.45 \cdot (z/0.1)^{-3/4}. \qquad (7.15)$$

At the spray boundary $r = r_{\lim}(z) = z \cdot \tan\varphi$, which is determined by the remotest (from the spray axis) trajectories of droplets forming with axis the angle $\varphi = 32.5°$, which is half the spray root angle, the values of $C_d(r_{\lim}, z)$ are obviously equal to about half (smaller by 1.8 times than) the value of $C_d(0, z)$ on the spray axis. This was taken into account for further calculations in the expression

$$C_d(r, z) = C_d(0, z) \cdot (0.45\exp(-6r/r_{\lim}(z)) + 0.55). \qquad (7.16)$$

In numerical simulation of the spray, before representing the differential equations by their difference analogs, it is expedient to pass to dimensionless variables, dividing the values of coordinates r and z by the initial (minimal) radius $r_0 = z_0 \cdot \text{tg} \varphi$ of the spray in the computation domain ($z_0 = 100$ mm from the output nozzle opening, $\varphi = 32.5°$ is the angle of spray cone); velocities w, u, V, and c_0 by initial velocity u_0 of droplets (liquid jets); and ρ by density ρ_0 of the resting gas far from the spray and t by $t_0 = r_0/u_0$. In this case, the forms of Eqs. (7.4–7.9) don't change, and the terms on the right-hand sides of the equations acquire corresponding additional coefficients.

Passing from differential Eqs. (7.4–7.9) to their difference analogs using representations (7.1) and (7.11–7.16) on the rectangular spatial grid (i, j), as in [4, 10], we used the explicit two-step Lax–Wendroff difference scheme [12]. In this scheme, intermediate values of dependent variables are determined for each temporal layer at the first (auxiliary) step of calculation for $t = (n + 1/2)\Delta t$ using the Lax scheme with half-step $\Delta t/2$. The values of quantities on the previous layer are averaged over four nearest nodes. At the second (main) step of calculation with total time step Δt, the resultant intermediate values of quantities are used in the expressions approximating spatial derivatives, and new values of variables are determined. Then, the cycle is repeated.

The Lax–Wendroff scheme is found to be centered in time [12]; for this reason, numerical effects of viscosity and diffusion in it are much weaker than in the one-step Lax scheme, which makes it possible to obtain the velocity profiles for each phase, which are closer to their true counterparts. To ensure the stability of the difference scheme, it is necessary to satisfy the Courant–Friedrichs–Lewy condition [12], which has the following form for identical spatial steps $\Delta z = \Delta r$ of the grid:

$$\Delta t \leq \frac{\Delta z}{\sqrt{2(c^2 + w_z^2 + w_r^2)}} \qquad (7.17)$$

The proposed model is advantageous because its difference equations make it possible to calculate variables using a simple explicit scheme. One of the difficulties encountered in constructing the numerical model is the specification of appropriate boundary conditions that preserve the stability of the difference scheme. In the boundary nodes of the domain of calculation, the difference scheme inevitably has a form that differs from that at the inner points; in these nodes of grid, spatial derivatives can be approximated using one-sided rather than two-sided differences. In addition, one should be in mind that radial velocities $w_r = u_r = 0$ of the phases on the symmetry axis (for $r = i \cdot \Delta r = 0$); the derivatives of some variables with respect to r can also vanish at these points.

At the upper (inlet) boundary of the domain of calculations ($j = 0$), it is necessary (taking into account experimental data) to specify the porosity profile, say, of triangular form $\alpha(r, z_0) = 3(r_n/r_0)^2(1 - r/r_0)$, r_n, being the radius of the outlet aperture of the sprayer.

At the upper boundary, the radial profiles $u_z(r, z_0)$ and $u_r(r, z_0)$ of liquid velocity components must be also specified. The former profile can be rectangular, trapezoidal, or more complicated. It is expedient to specify the second profile by the linear function of radius, $u_r(r, z_0) = u_z(r, z_0)\, r/z_0$, taking into account of the swirl form of the liquid flow from the sprayer nozzle. Instead of the experimental data, we can use the results of previous calculations for the region closest to the sprayer.

In calculating the free two-phase jet of spray on its lateral (external) boundary, the gas density can be determined using the Bernoulli equation

$$\rho = \rho_0 \left(1 - \frac{w_z^2 + w_r^2}{2c_0^2}\right). \tag{7.18}$$

To calculate the two-phase flow in the spray apparatus, the conditions of vanishing of the gas velocity components ($w_z = w_r = 0$) are specified at the lateral boundary (wall of the cylindrical apparatus).

7.2.2 Results of Calculating the Free Spray

The above-described algorithm was realized using the Delphi program package to calculate the two-phase flow in an axisymmetric spray produced by centrifugal-spray injector with a nozzle diameter of 2 mm. During code debugging, the difference form of the boundary condition was varied, while in the numerical experiment, we varied the form of dependences (7.12) and (7.13) for stress of turbulent friction of the gas and the drag coefficient of droplets using formulas (7.2), (7.3), and (7.16), as well as $C_d = 0.17$ and 0.1.

The unit of dimensionless spatial scale of the grid was the spray radius at the upper boundary $r_0 = z_0 \cdot \tan\varphi$ of the domain of calculations ($\varphi = 32.5°$ is half the root angle of the spray); for the unit of the velocity scale, the initial velocity $u_0 = 0.75$ $(2P_1/p_1)^{1/2}$ of liquid flow from the sprayer nozzle was used. The calculations were performed for water pressure $P_1 = 5 \times 10^5$ Pa at the sprayer; the measured droplet diameter was $d = 0.14$ mm. In our calculations, the value of $z_0 = 100$ mm was used.

The dimensionless grid pitch was defined as $h = \Delta r = \Delta z = 1/N$, $N = 16$; the size of the spatial region was $r_{max} = \max(i) \cdot h$, $z_{max} - z_0 = \max(j) \cdot h$. The size of the rectangular grid was varied to $\max(i) = \max(j) = 200$ and ensured sufficient approximation of the difference scheme.

It has been established that the (quasi-)stationary state of the flow which is interest for us in the present problem was attained as a result of evolution of the time-dependent solution. The dimensionless time required for this is approximately 15–20

Fig. 7.2 Variations in velocities of droplets (uz) and gas (wz) on spray axis; curves were calculated taking into account drag crisis by formula (7.16); uz_1 and wz_1 curves were calculated using formula (7.3); $j = -25$ is position of the outlet aperture of the nozzle [8]

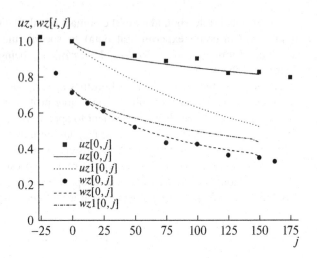

Fig. 7.3 Radial profiles of the axial velocity of the gas at different distances $z = (100 + 4j)$ mm from the nozzle; curves are calculation results, and symbols represent experimental data [8]

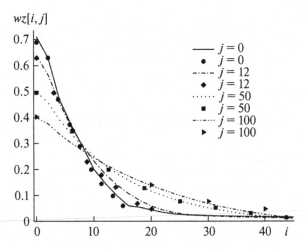

times more than the characteristic time $t_s = (z_{max} - z_0)/u_0$, at which the droplets cross the calculations domain from the upper to lower boundary without accounting their deceleration in the gas.

Figures 7.2, 7.3, and 7.4 show the results of calculations of a free axisymmetric spray based on the proposed model.

Figure 7.2 shows the results of the calculation and experimental data [3] for the velocities of phases at the spray axis for the values of C_d calculated using formulas (7.16) and (7.3). In the former case, the agreement between the calculation and experiment is obvious. In the latter case, droplets are decelerated, and the gas is accelerated more intensely than in the experiment.

Fig. 7.4 Radial profiles of axial velocity of droplets at different distances $z = (100 + 4j)$ mm from the nozzle; curves are calculation results, and symbols represent experimental data [8]

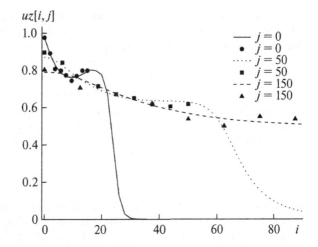

Figure 7.3 shows the axial velocity profiles for the gas in the spray at distances $z = 100, 150, 300,$ and 500 mm from the nozzle. Symbols show experimental results from [3], and the curves are obtained as a result of calculations.

Figure 7.4 shows analogous profiles for droplets at distances $z = 100, 300,$ and 700 mm from the sprayer. The curve for $j = 0$ and $z = 100$ mm describes the dependence that approximates the experimental data [3].

Figures 7.2, 7.3, and 7.4 demonstrate satisfactory agreement between the calculations results of a free two-phase flow of spray with accounting early crisis of drag for droplets and experimental results for a considerable length of the spray region of 100–700 mm from the nozzle.

7.2.3 Calculation of the Two-Phase Flow in a Spray Apparatus

Among the spray apparatus used in chemical technologies, two types can be distinguished, i.e., injectors and ejectors [13]. Each of them contains cylindrical chamber for phases mixing that is coaxial with the sprayer and a separator tank for their separation. The difference between these apparatus is that the gas flow rate in the injector is limited by the valve at the apparatus outlet, while the valve in the ejector is installed at the inlet.

The presence of the cylindrical chamber limits the two-phase flow on radius and height. The inner surface of its wall is the lateral boundary $i = n$ of the computation domain, on which both gas velocity components vanish: $wz(i, j) = uz(i, j) = 0$. Droplets freely precipitate on the wall. The height H of the mixing chamber is connected with the position of the lower boundary $j = nz$ of the computation domain by the relation $H = (100 + 4nz)$ mm.

In calculations, the gas pressure drop P between the lower and upper cross sections of the mixing chamber (computation domain boundaries) was specified. The volume gas flow rate Q in the apparatus was determined from the calculated values of the gas axial velocity. The variants of calculation of the apparatus differed in the gas pressure drop P and radius R_{APP} of the mixing chamber of the apparatus.

The results of calculations were used to determine the dependence $P(Q)$ known as the hydraulic characteristic of the apparatus and maximal values P_m and Q_m and their dependence on R_{APP}, as well as the hydraulic efficiency of the apparatus in the form of function

$$\text{eff}(Q) = P \cdot Q/(P_1 \cdot Q_1) \tag{7.19}$$

from gas flow rate Q. Here P_1 and Q_1 are the pressure drop and the flow rate of the liquid at the sprayer. In calculations, the following experimental data for these quantities were used: $P_1 = 5 \cdot 10^5$ Pa and $Q_1 = 0.745 \cdot 10^{-4}$ m^3.

The calculations results are presented in Figs. 7.5, 7.6, 7.7, 7.8, and 7.9.

Figure 7.5 shows the axial velocity profiles of droplets and gas in the spray apparatus at various distances z from the nozzle. Figure 7.6 shows the profiles of the radial gas velocity in different radial cross sections of the same apparatus for the same gas pressure drop in it ($P = 14$ Pa). The negative values of the wz and wr projections of the gas velocity onto the coordinate axes near the apparatus wall ($i = 35$) indicate the reverse vortex flow of the gas near the upper ($j = 0$) inlet cross section of the apparatus.

Figure 7.7 shows the calculated dependences of the maximal values of the pressure drop P_m and volume gas flow rate Q_m in the apparatus from its radius R_{APP}.

It can be seen that, upon an appreciable increase in the apparatus cross-sectional area (by four times), the maximal gas pressure drop P_m across it changes insignificantly (by 7%). At the same time, the maximal gas flow rate Q_m in the apparatus

Fig. 7.5 Calculated axial velocity profiles of droplets (top) and gas (bottom) at different distances $z = (100 + 4j)$ mm from nozzle in spraying apparatus of the radius $R_{APP} = 140$ mm and height $H = 800$ mm [8]

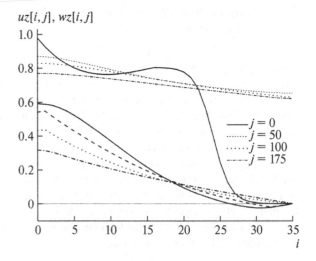

Fig. 7.6 Calculated radial velocity profile for the gas at distances $z = (100 + 4j)$ mm from the nozzle in the apparatus of radius $R_{APP} = 140$ mm and height $H = 800$ mm [8]

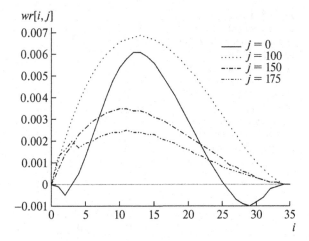

Fig. 7.7 Calculated dependences of maximal values of (■) pressure drop P_m of the gas in the apparatus and (●) gas volume flow rate Q_m from radius R_{APP} of apparatus [8]

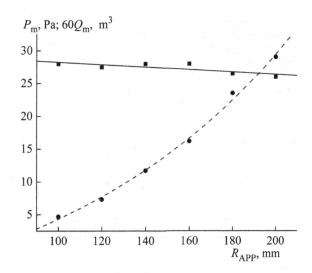

changes significantly (by six times), approximately following the power law $Q_m = 0.08 \, (10R_{APP})^{2.6}$, if Q_m is measured in cubic meters and R_{APP} in meters.

It can be seen that, upon an appreciable increase in the apparatus cross-sectional area (by four times), the maximal gas pressure drop P_m across it changes insignificantly (by 7%). At the same time, the maximal gas flow rate Q_m in the apparatus changes significantly (by six times), approximately following the power law $Q_m = 0.08(10R_{APP})^{2.6}$, if Q_m is measured in cubic meters and R_{APP} in meters.

Figure 7.8 shows the calculated dependence of the relative pressure drop $p = P/P_m$ of the gas in the apparatus on the relative volume flow rate of gas $q = Q/Q_m$ through the apparatus.

Fig. 7.8 Calculated dependences of relative pressure drop $p = P/P_m$ of gas in apparatus on relative volume flow rate $q = Q/Q_m$ of gas through apparatus for two radii R_{APP}: (■) 140 and (●) 160 mm; P_m and Q_m are the maximal values of P and Q. Curve was calculated by using dependence (7.20) [8]

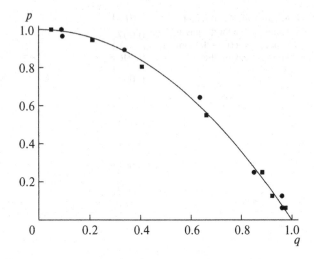

Fig. 7.9 Calculated dependences of relative efficiency $e = eff/eff_m$ of apparatus from relative volume flow rate q of gas through apparatus for two radii R_{APP}: (■) 140 and (●) 160 mm. Curve was calculated by using dependence (7.21) [8]

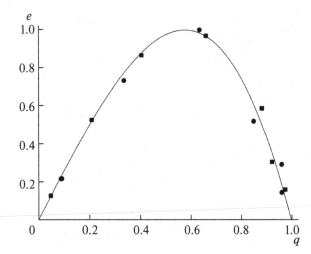

Figure 7.9 shows the calculated dependences of the relative hydraulic efficiency $e = eff/eff_m$ of the apparatus from the relative volume flow rate of the gas q through the apparatus. For reference, it should be noted that $eff_m = 0.055$ for $R_{APP} = 140$ mm and $eff_m = 0.083$ for $R_{APP} = 160$ mm.

It is interesting to note that the $p(q)$ and $e(q)$ dependences obtained in [13] from simple theoretical premises, which have the simple form

$$p = 1 - q^2, \tag{7.20}$$

$$e = 3^{3/2}/2 \cdot q(1 - q^2) \tag{7.21}$$

and coincide with the experimental results obtained on spray apparatuses, were also found to be in conformity with the results of calculations based on a more rigorous model of the spray process described above. This is obvious from Figs. 7.8 and 7.9.

Formula (7.21) implies that the maximal value of hydraulic efficiency (eff_m) of the spraying apparatus is attained at the optimal gas flow rate Q_{opt}, which satisfies the condition

$$q_{opt} = Q_{opt}/Q_m = 3^{-1/2}. \qquad (7.22)$$

7.2.4 Conclusions About Two-Phase Flow Produced by a Nozzle

In this study, the model of the spray produced by a nozzle, which was proposed earlier in [4, 10], has been improved. A comparison of the results of calculations of the two-phase flow based on this model with the experimental data leads to the conclusion that the numerical model described here makes it possible to calculate the turbulent two-phase flow with an admissible degree of accuracy.

The distinguishing features and advantages of this model are that it is applicable to a compressible continuous dispersion medium, as well as for an almost incompressible flow; it can be used to calculate nonstationary flows, as well as (quasi-) stationary state. The latter can be obtained from the former states as a result of evolution. The difference equations of the given model make it possible to calculate all variables using a simple explicit scheme.

As compared to earlier publications of the author [4, 10], new results have been obtained in this study. In particular, the free spray with a height of up to 1 m was calculated in one run and not only in several runs on sequential regions of a smaller height. The results of new calculations for the velocities of the phases are in good agreement with experiment.

In addition to the free spray, we have calculated the two-phase flow of droplets and a gas in a cylindrical spray apparatus. The dependences of maximal values of gas volume flow rate Q_m and pressure drop P_m in the apparatus from its diameter have been established. And also the form of the dependences of hydraulic characteristic $P(Q)$ of the apparatus and its hydraulic efficiency $eff(Q)$ from gas flow rate Q in the apparatus have been calculated. It has been shown that the peak of the hydraulic efficiency is attained under condition (7.22).

Taking into account the analogy between interfacial transfer of momentum, mass, and heat in a turbulent two-phase flow [1, 2, 11], we can suppose that the maximal efficiency of the heat and mass transfer in the spraying apparatus is attained at the same condition (7.22).

The results presented in Sects. 7.1 and 7.2 were previously published in article [8].

7.3   Calculation of Interphase Mass Transfer in a Spray Flow Produced by a Nozzle

The above-described mathematical model of two-phase flow of liquid spray produced with a nozzle was also used in this part of work. Axial profiles of the gas impurity concentrations in a free spray have been calculated and presented, as well as radial profiles of gas impurity concentrations in a two-phase flow through a cylindrical apparatus, taking into account early crisis of drag for droplets and the mass transfer crisis. Dependences of gas flow rate, concentration of the gas admixture at the outlet of the apparatus, and the amount of impurity absorbed by the liquid from the height and area of the apparatus cross section were established by calculation.

Often the process of interphase mass transfer is accompanied by parallel and simultaneous heat exchange, which complicates their modeling and calculation. In the technology of wet cleaning of air from harmful gas impurities [14], for example, sulfur dioxide SO_2, the interfacial exchange process can be considered without taking heat exchange into account. Modeling of this kind of mass exchange processes associated with the absorption of a gas admixture (in the air) by the drops of a sprayed liquid (water) was the goal of this part of researches.

7.3.1   Simulation of Mass Transfer in a Free Spray Flow

Equations (7.4–7.16) of hydrodynamics of a two-phase flow in spray of a nozzle, which form the basis of the model under consideration, are described in detail in Sect. 7.2.1.

To calculate the mass exchange of phases (without taking into account their heat exchange), this system of equations must be supplemented by relationships that take into account the convective mass transfer of the impurity gaseous component in a spray flow from the gas into droplets of liquid.

Taking into account the analogy of the phenomena of interphase heat and mass transfer [11] for the density of impurity mass flux from gas into a separate drop, we can write equation which is analogous to (6.40)

$$j = k(C - C_1)\rho. \tag{7.23}$$

This equation is analogous to the Newton–Richman equation for the heat flow. Also, by analogy with the heat transfer coefficient, we can determine the mass transfer coefficient by using the relationship

$$k = \mathrm{Nu}_D \cdot D/d. \tag{7.24}$$

The diffusion Nusselt number (sometimes called the Schmidt number) can be determined using the Ranz–Marshall formula proposed for heat exchange of a sphere with a liquid flow

$$\mathrm{Nu}_D = 2 + 0.6\mathrm{Re}^{1/2}\mathrm{Pr}_D^{1/3}. \tag{7.25}$$

Here and below, $C = C(r, z)$ is the concentration of the gas admixture in the air far from the droplet, C_1 is the same near its surface, D is the diffusion coefficient of the impurity (for SO_2 in air, $D = 2.45 \cdot 10^{-5}$ m^2/s), and $\mathrm{Pr}_D = \mu/(\rho D) = 0.637$ is the diffusion Prandtl number.

Without taking into account the crisis of mass exchange analogous to the heat exchange crisis [15, 16], formula (7.25) gives $\mathrm{Nu}_D > 2$. To take into account the mass exchange crisis, we need to use the value $\mathrm{Nu}_D = 2$.

To simplify the model, the concentration of easily absorbed (as in case of chemisorption) gas impurity by droplets liquid at their surface can be assumed to be $C_1 = 0$. This is the external task of mass transfer of phases.

Then the mass impurity flow from the gas into all drops per unit volume of spray flow, with allowance for (7.23), can be represented in the form

$$J = \frac{j4S\alpha}{V_d} = \frac{6j\alpha}{d} = \frac{6k\rho\alpha C}{d}. \tag{7.26}$$

Now the equation of convective-diffusion transfer of gaseous impurity component of spray stream from gas into drops of liquid can be written in the form

$$\frac{\partial C}{\partial t} + w_z \cdot \frac{\partial C}{\partial z} + w_r \cdot \frac{\partial C}{\partial r} = D \cdot \Delta C - \frac{J}{\rho}. \tag{7.27}$$

Equation (7.27), together with relations (7.24–7.26), completes the system of Eqs. (7.4–7.16) and allows one to calculate the change in the concentration of the gas admixture $C = C(r, z)$ in a two-phase spray flow of the nozzle taking into account the crises of drag of drops and interphases mass exchange and without them.

In the numerical model of the spray of a spray, before presenting the differential equations with their difference analogs, it is advisable to proceed to dimensionless variables, dividing them by the corresponding characteristic values, as indicated in Sect. 7.2.1.

In the numerical model of the spray flow, before presenting the differential equations by their difference analogs, it is advisable to proceed to dimensionless variables, dividing them by the characteristically values, as indicated in Sect. 7.2.1. The value of $C(r, z)$ must be divided by the initial impurity concentration C_0 in the stream of yet non-cleaned gas.

The forms of Eqs. (7.4–7.9), (7.27) don't change, and the corresponding additional coefficients appear in the right-hand sides of the equations.

As in works [4, 8, 10] and Sect. 7.2, when passing from the differential Eqs. (7.4–7.9) and (7.27) to their difference analogs, taking into account the representations (7.1) and (7.11–7.16) on a rectangular spatial grid (i, j), to approximate the convective terms, we used the explicit two-step difference scheme of Lax–Wendroff [12]. For its stability, it is necessary to fulfill the Courant–Friedrichs–Levy condition (7.17).

In combination with the Lax–Wendroff scheme, the approximation of the diffusion term in Eq. (7.21) was carried out according to an explicit scheme of the first order of accuracy [12, p. 107], for which, in the case of a two-dimensional net, the stability condition has the form

$$\Delta t \leq \frac{(\Delta z)^2}{4D}. \tag{7.28}$$

To ensure the stability of the difference scheme as a whole, it is necessary to simultaneously fulfill both conditions (7.17) and (7.28) [12, p. 290], of which the first was stronger in this case.

The advantages and disadvantages of the described difference scheme, as well as the boundary and initial conditions of the model, were specified in Sect. 7.2.1.

In addition to this, we note that when calculating the free spray flow on its lateral (external) boundary, we can use the initial value $C_0 = 1$ to determine the dimensionless impurity concentration in the gas.

When calculating the two-phase flow in spray cylindrical apparatus at the lateral boundary (the wall of the apparatus) for the impurity concentration, we can assume the condition $C(R_{APP}, z) = 0$ for $z \geq z_w = R_{APP}/tg\varphi$ due to the wetting of the wall by the droplet liquid that readily absorbs the impurity.

7.3.2 Results of Calculating the Mass Transfer of Phases in a Free Spray Flow

The algorithm described above was implemented using Delphi software to calculate the interfacial mass transfer in a two-phase flow of an axisymmetric spray of a centrifugal-jet injector with a nozzle diameter of 2 mm.

In the numerical experiment, the dependence (7.12) was used for the stress of turbulent friction of the gas, and formulas (7.15) and (7.16) were used for the drag coefficient of drops.

The calculation parameters such as r_0, z_0, φ, u_0, P_1, d, Δr, Δz, r_{max}, z_{max}, h, max (i), and max(j) were adopted same as in Sect. 7.2.2.

The quasi steady state of the flow, which we were interested in, was achieved as a result of the evolution of the no stationary solution over the grid time, approximately 15–20 times greater than the characteristic time $(z_{max} - z_0)/u_0$, for which the drops

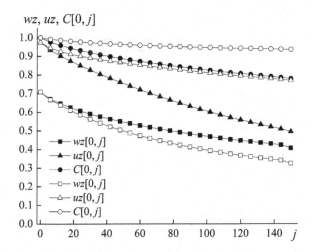

Fig. 7.10 Calculated dependences for the velocities of gas $wz[0,j]$ and drops $uz[0,j]$ as well as for the concentration $C[0,j]$ of the gas impurity on the axis of free spray flow; dark symbols correspond to calculations by using formulas (7.3) and (7.25) without taking into account the early crises of drag of drops and mass transfer ($Nu_D > 2$); bright symbols correspond to calculation taking into account the drag crisis by formula (7.16) and the mass exchange crisis at $Nu_D = 2$ [17]

could have flown from the upper to the lower boundary of the calculated region without taking into account their braking in the gas.

Figures 7.10 and 7.11 show the results of the calculation by using the proposed model of interfacial mass transfer in an axisymmetric free spray flow of nozzle.

Figure 7.10 shows the axial profiles of the phases velocities and concentration C $[0,j]$ of the impurity on the axis of free spray flow without and with taking into account the crises of early drag of droplets and interfacial mass transfer. Evidently, the drag crisis has a marked effect on the velocity profiles of each phase, the gas moves more slowly, and the droplets are faster due to a weaker interphase exchange by momentum.

The early crisis of drag of droplets (without taking into account the crisis of mass exchange, when according to formula (7.25) $Nu_D > 2$) is small affected on the interphase mass transfer, so, at distance $z = 700$ mm from the nozzle (at $j = 150$), the calculated impurity concentration, taking into account the drag crisis, was only 2% greater than without it (in Fig. 7.2 is not shown). With an additional account of the mass exchange crisis, when the values of $Nu_D = 2$, the intensity of the interfacial mass transfer decreases significantly, as seen in Fig. 7.10.

Figure 7.11 shows the radial profiles of the concentration $C = C\,[i,j]$ of the impurity at different distances from the nozzle.

Obviously, the crisis of drag of drops significantly affects both the position of the minimum impurity concentration along the radius of the apparatus and its numerical value at this minimum. And this influence grows with the distance z from the nozzle.

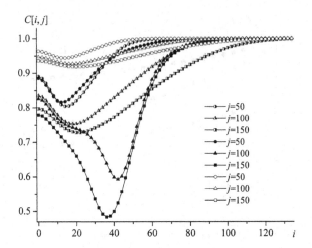

Fig. 7.11 Calculated radial profiles of the concentration $C[i, j]$ of the gas admixture at different distances $z = (100 + 4j)$ mm from the nozzle in the free flow of spray; dark symbols correspond to calculation using formulas (7.3) and (7.25) without taking into account the crises of drag of drops and mass transfer ($\mathrm{Nu}_D > 2$); bright symbols correspond to calculation taking into account the crisis of drag according to the formula (7.16) and the mass exchange crisis (with $\mathrm{Nu}_D = 2$); half dark and half bright symbols correspond to calculation taking into account the drag crisis, but without taking into account the mass exchange crisis [17]

The additional accounting of the crisis of mass exchange does not affect the position of the minimum, but it has a noticeable effect on the minimum and other values of the impurity concentration in each of these cross sections of the apparatus.

7.3.3   Calculation of Interfacial Mass Transfer in a Spraying Apparatus

The spraying apparatuses used in chemical technologies can be conditionally divided into two types: injectors and ejectors [13]. In each of them, there are a nozzle 1 forming a spray flow 2, the cylindrical chamber 3 coaxial with the nozzle for mixing of phases, and a tank 4 to separate them (Fig. 7.12). The axis of the nozzle and the chamber are common and vertical.

The difference of apparatuses is that the mixing chamber of the injector is open from above; the gas flow is limited by the valve 5 at the gas outlet from the apparatus. But in the ejector, the mixing chamber, inside which the nozzle is located, is closed from the top; the gas flow is limited by the valve 5 at the gas inlet through the top cover of the apparatus (it isn't shown in Fig. 7.12). The liquid from the layer 6 accumulated in the separator tank is discharged through the valve 7.

The cylindrical chamber for mixing of phases limits the two-phase flow over the radius ($r < R_{\mathrm{APP}}$) and height H. Last is related to the position of the lower boundary

Fig. 7.12 Scheme of a spraying injector apparatus for cleaning air from a gas admixture, absorbed by liquid droplets [17]

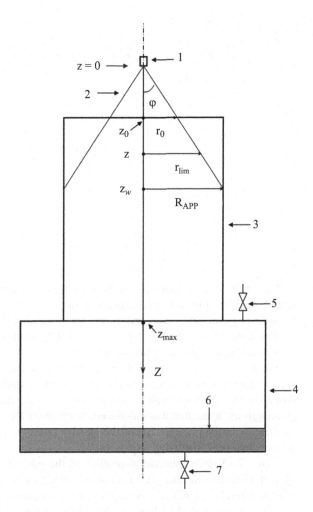

$j = nz$ of the computational domain by the formula $H = z_{max} = (z_0 + h \cdot nz)$ mm. The inner surface of chamber wall is the lateral boundary $i = n = R_{APP}/h$ of the computational domain, on which both components of the gas velocity vanish: $w_r(n, j) = w_z(n, j) = 0$.

The droplets reaching the wall drop out onto it, wetting the inner surface of the chamber below the coordinate $z_w = R_{APP}/tg\varphi$. For the concentration of the impurity gas component, which is easily absorbed by the droplet liquid on the side wall of the apparatus, two types of boundary conditions were tested:

1. $C(R_{APP}, z) = 1$ for $z < z_w$ and $C(R_{APP}, z) = 0$ for $z \geq z_w$
2. $C(R_{APP}, z) = 0$ for all possible values of z under the assumption that air vortices can transfer droplets on the apparatus wall above the level z_w

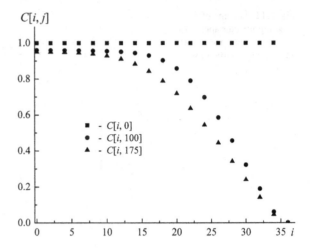

Fig. 7.13 Calculated radial profiles of the concentration $C[i, j]$ of the gas admixture at various distances $z = (100 + 4j)$ mm from the nozzle in the injector apparatus; $R_{APP} = 140$ mm, $H = 800$ mm [17]

The results of the calculations in these two cases differed slightly. But in the basic series of calculations, the first variant was used.

In the proposed spray flow model, the computational domain $0 < j < nz$ is entirely located in the phases mixing chamber (Fig. 7.12), so the calculation algorithm does not depend on the type of spraying apparatus.

In calculations, the pressure drop ΔP of the gas was set between the lower and upper cross sections of the mixing chamber – the boundaries of the calculated region. Based on the calculated values of the gas axial velocity, the volume flow of gas Q through the apparatus was calculated. Variants of the device calculation differed by the differential pressure ΔP of the gas, the radius R_{APP}, and the height H of the mixing chamber of spraying apparatus.

Figure 7.13 shows the radial profiles of the gas impurity concentration in the apparatus, calculated taking into account both crises: hydrodynamic resistance and mass transfer. Obviously, the absorption of the gaseous impurity near the axis of the apparatus is much weaker than the absorption near its wall.

Figure 7.14 shows the results of the calculation of the dependences for the average impurity concentration $<C>$ (over the output cross section of the mixing chamber) and the amount $Q(1 - <C>)$ of the gas admixture absorbed by the liquid in the apparatus versus the volume flow rate Q of gas. Obviously, the average concentration of impurities $<C>$ slightly increases (by about 20%) at increasing of gas consumption Q by an order of magnitude. The amount of impurity absorbed by the liquid increases almost the same as the gas consumption.

The weak growth of $<C>$ with increasing Q confirms the assumption made in [8] that the maximum mass transfer efficiency in the spraying apparatus may be achieved with a gas flow close to the optimal Q_{opt} satisfying the condition (7.22).

Figure 7.15 shows the calculated dependences of the maximum flow rate Q_m of the gas (i.e., at a small difference $\Delta P = 0.7$ Pa of the gas pressure on the apparatus), as well as of the quantities $<C>$ and $Q_m(1 - <C>)$ from the cross-sectional area S of

Fig. 7.14 Calculated dependences for the average impurity concentration $<C>$ and as well as for tenfold quantity $Q(1 - <C>)$ of the gas impurity absorbed by the liquid (in the drops and in the film flowing on the apparatus wall) versus the gas flow rate Q; $R_{APP} = 140$ mm, $H = 1100$ mm [17]

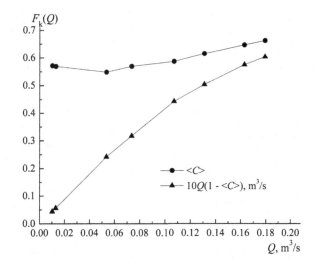

Fig. 7.15 Calculated maximum flow rate Q_m of the gas through the apparatus as well as the average impurity concentration $<C>$ and the tenfold maximum amount $Q_m(1 - <C>)$ of the gas admixture absorbed by the liquid versus the area S of the apparatus cross section; $H = 1100$ mm, gas pressure drop on the apparatus $\Delta P = 0.7$ Pa [17]

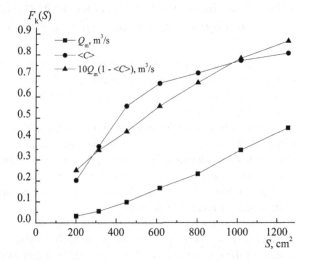

the mixing chamber. It is obvious that all values increase with increasing S. Growth $<C>$ means that the intensity of the mass transfer of phases and the degree of air purification are reduced, although the total amount of impurities extracted from it is increased due to an even greater increase in the flow rate Q_m of the gas.

Figure 7.16 shows the calculated dependences for the values $<C>$, $2Q$ and $10Q$ $(1 - <C>)$ from the height H of apparatus. It is obvious that all quantities with increasing H vary approximately linearly. With an increase in H by a factor of 2.7, a nearly twofold decrease in the gas flow rate Q is compensated by an almost threefold increase in the fraction $(1 - <C>)$ of the impurity absorbed by the liquid. As a result, the total amount of absorbed impurity $Q(1 - <C>)$ increases by 1.5 times.

Fig. 7.16 Calculated average impurity concentration $<C>$, the doubled flow rate Q of the gas through the apparatus, and the tenfold amount Q $(1 - <C>)$ of the gas impurity absorbed by the liquid versus the height H of the apparatus; $R_{APP} = 140$ mm, gas pressure drop $\Delta P = 7$ Pa; the solid line shows the linear approximation of the calculated dependence $<C>$ versus H [17]

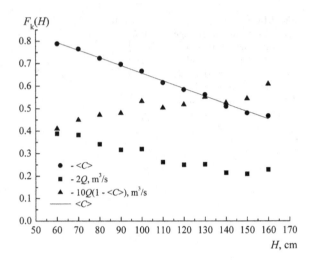

7.3.4 Conclusions About Interphase Mass Transfer in a Spray Flow Produced by a Nozzle

The model of the spray flow produced by a nozzle proposed earlier in [4, 8, 10], taking into account the early crisis of drag of drops, is developed in this paper with additional accounting of the mass transfer crisis from gas to drops analogous to the crisis of heat transfer [15, 16].

In comparison with the previous works of the author [4, 8, 10], new results were obtained in this paper. In particular, not only axial and radial profiles of phases velocities were calculated in the free spray flow of height H up to 0.7 m but also the distribution of the $C(r, z)$ concentration of the gaseous impurity in the spray flow with accounting of the crises of drag for droplets and interphase mass transfer and without taking them into account.

In addition to the free flow of spray, the interphase mass transfer was also calculated in a flow of gas and drops through a cylindrical apparatus. The dependences of the concentration $<C>$ of the gas impurity at the output of the apparatus and the amount $Q(1 - <C>)$ of the gas admixture absorbed by the liquid have been determined versus the gas volume flow rate Q through the apparatus, as well as the dependences of the three indicated quantities on the cross-sectional area S and the height H of apparatus.

The presented numerical model makes it possible to calculate the dependence of the mode characteristics Q, $<C>$, $Q(1 - <C>)$ of the spray apparatus on its design parameters S and H and the pressure drop ΔP of the gas on it.

According to Figs. 7.14, 7.15, and 7.16 due to the mass transfer crisis, the degree of gas purification from the impurity $(1 - <C>)$ is small and does not exceed 0.5. Therefore, it is worth considering the possibility of using several stages of air purification from the impurity.

The results of Sect. 7.3 are presented in work [17].

References

1. Nigmatulin, R. I. (1987). *Dynamics of multiphase media.* Moscow: Nauka. Hemisphere, New York, 1991.
2. Abramovich, G. N. (1984). *Theory of turbulent jets.* Moscow: Nauka.
3. Simakov, N. N. (2004). Crisis of Hydrodynamic Drag of Drops in the Two-Phase Turbulent Flow of a Spray Produced by a Mechanical Nozzle at Transition Reynolds Numbers. *Technical Physics, 49,* 188.
4. Simakov, N. N., & Simakov, A. N. (2005). Anomaly of gas drag force on liquid droplets in a turbulent two-phase flow produced by a mechanical jet sprayer at intermediate Reynolds numbers. *Journal of Applied Physics, 97,* 114901.
5. Torobin, L. B., & Gauvin, W. H. (1959). *Canadian Journal of Chemical Engineering, 37*(4), 129.
6. Schlichting, H. (1955). *Boundary layer theory.* New York: McGraw-Hill. Nauka, Moscow, 1974.
7. Simakov, N. N. (2010). Experimental Verification of the Early Crisis of Drag Using a Single Sphere. *Technical Physics, 55,* 913.
8. Simakov, N. N. (2013). Calculation of the flow about a sphere and the drag of the sphere under laminar and strongly turbulent conditions. *Technical Physics, 58,* 481.
9. Simakov, N. N. (2017). Numerical Simulation of the Two-Phase Flow Produced by Spraying a Liquid by a Nozzle. *Technical Physics, 62,* 1006. https://doi.org/10.1134/s1063784217070222.
10. Simakov, N. N. (2002). Numerical simulation of a two-phase flow in the spray stream produced by the nozzle. *Izvestiia Vysshykh Uchebnykh Zavedenii. Khimiya and KhimicheskayaTekhnologiya (News of universities. Chemistry and chemical technology), 45* (7), 125.
11. Loitsyanskii, L. G. (1978). *Mechanics of liquids and gases.* Moscow: Nauka. Pergamon, Oxford, 1966.
12. Potter, D. E. (1973). *Computational physics.* London: Wiley. Mir, Moscow, 1975.
13. Simakov, N. N. (2003). Dissertation, Yaroslavl State Technical Univ., Yaroslavl.
14. Simakov, N. N. (2016). Calculation of the Drag and Heat Transfer from a Sphere in the Gas Flow in a Cylindrical Channel. *Technical Physics, 61,* 1312.
15. Simakov, N. N. (2016). Calculations of the Flow Resistance and Heat Emission of a Sphere in the Laminar and High-turbulent Gas Flows. *Technical Physics, 61,* 1806.
16. Straus, V. (1981). *Promyshlennaya ochistka gazov (Industrial gas purification).* Moscow: Khimiya.
17. Simakov, N. N. (2017, in press). *Raschet mezhfaznogo massoobmena v fakele raspyla forsunki (Calculation of inter-phase mass exchange in a spray flow of a nozzle).* Zh. Tekh. Fiz. (Tech. Phys.).

Index

© Springer Nature Switzerland AG 2020
N. N. Simakov, *Liquid Spray from Nozzles*, Innovation and Discovery in Russian
Science and Engineering, https://doi.org/10.1007/978-3-030-12446-5

Printed in the United States
By Bookmasters